VENOMOUS AND POISONOUS ANIMALS

VENOMOUS AND POISONOUS ANIMALS

Anders Edstrom

KRIEGER PUBLISHING COMPANY
MALABAR, FLORIDA
1992

First English Edition 1992

Printed and Published by
KRIEGER PUBLISHING COMPANY
KRIEGER DRIVE
MALABAR, FLORIDA 32950

Printed in the United States of America.

Library of Congress Cataloging-In-Publication Data
Edström, Anders, 1938–
Venomous and poisonous animals / Anders Edstrom.—1st English ed.
p. cm.
Includes bibliographical references and index.
ISBN 0-89464-627-3 (alk. paper)
1. Poisonous animals. 2. Poisonous animals—Toxicology. 3. Venom. I. Title.
QL100.E37 1992
615.9'4—dc20 91-17454
CIP

10 9 8 7 6 5 4 3 2

Contents

1

Animal Venoms Have an Important Biological Function

In the marine biological expedition to the Gulf of Lower California, which John Steinbeck has described in his wonderful book, *The Log From The Sea Of Cortez*, he often returns to all the venomous organisms, which the sea seems to be swarming with. In one place he writes, "Caught against the rocks by the current was a very large pelagic coelenterate, in appearance like an anemone with long orange-pink tentacles, apparently not retractable. On picking him up we were badly stung. His nettle-cells were vicious, stinging even through the calluses of the palms, and hurting like a great many bee stings. At this entrance also we took several giant sea-hares, a number of clams, and one small specimen of the clam-like hacha. For hours afterwards the sting of the anemone remained. So very many things are poisonous and hurtful in these Gulf waters: urchins, sting-rays, morays, heart-urchins, this beastly anemone, and many more. One becomes very timid after a while."

To defend oneself, to avoid becoming a prey of other animals, is a prerequisite for life—for reproduction—which in turn requires energy and building materials, that is, food. In this simplified structure of life, some animals defend themselves and others obtain food by using venoms; chemical tools which during evolution have been formed into some of the most interesting chemical substances we know.

The existence of venomous animals within the animal kingdom appears to be a rule rather than an exception. An investigation of common coral reef invertebrates found at the north Great Barrier Reef outside Australia showed that 73% of 42 species studied were very toxic to fish. Using more sensitive toxicity tests still more venomous invertebrates would probably have been detected. Apparently, at the bottom of the sea an intense chemical warfare is being conducted. However, venomous or poisonous animals are not only found in the sea but are also widely distributed throughout the animal kingdom, ranging from the unicellular organisms to the mammals, for example, the platypus and the spiny anteater. In fact, poisonous organisms are so numerous that scientists are still many years away from analyzing all the toxic molecules found in such animals. Birds appear to be the only large and widespread group of animals, in which no venomous species have been discovered.

Among marine organisms the venoms generally fulfil important defense functions. It also occurs, that animals use their venoms in offense, to capture prey. This has, perhaps, attracted the most public interest, especially in relation to venomous snakes. Some animals, for instance scorpions, use their venom for both offense and defense. It is usual that venom used for hunting is produced by glands associated with the mouth region and is delivered to the victim by the teeth or other mouth parts. Defensively designed venoms are mostly associated with other parts of the body. In the dart-poison frog, venomous secretions are released from microscopic glands scattered in the skin. Venomous fishes have a variety of spines and stings on the body, often associated with the fins, by which toxins are injected (Figure 1). Defensive venoms often produce severe pain in man, although there are several exceptions and no general rules.

In some animals, the venoms are synthesized by highly specialized cells. Certain animals have developed a natural resistance to such toxins and are able to eat the producers, be they animals or

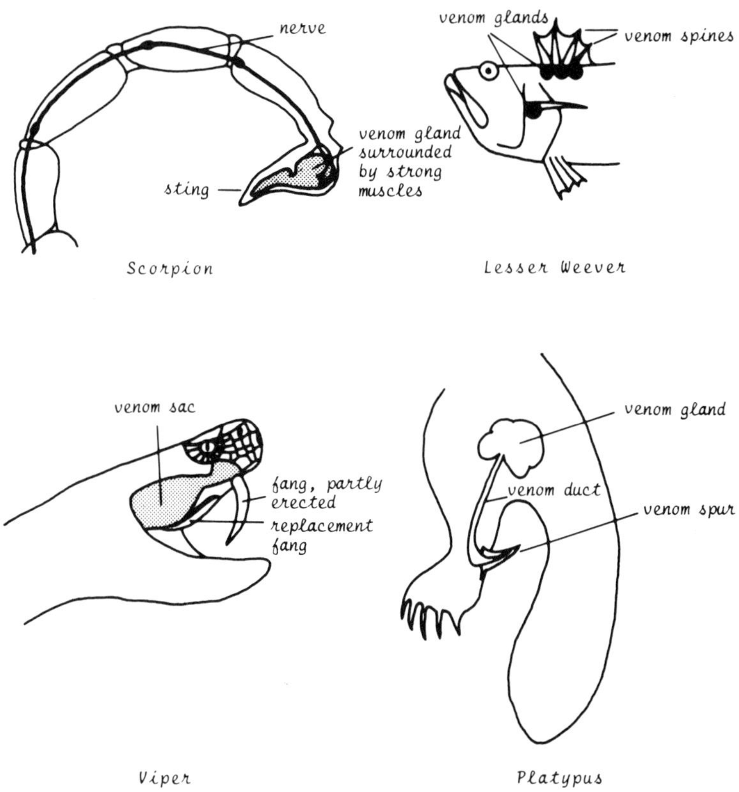

Figure 1. Examples of different venom apparatuses.

plants. In some cases they have learned to exploit the toxins for their own defense purposes through the ingestion of certain toxic organisms. This phenomenon is common among insects and marine organisms.

The terms "venomous" and "poisonous" are often used synonymously, but usually represent distinct meanings for the toxinologist. A venomous animal is equipped with a special apparatus for synthesizing and delivering its venom. In contrast, poisonous animals lack such devices but their tissues are in part or are all toxic for predators. In poisonous animals the toxins may or may not have a role in natural defense mechanisms. Where the word poisonous animal is used in this text, it refers to an organism, in which the toxin has some biological origin. It does not refer to animals poisoned by toxic substances synthesized by man, such as DDT, PCB, and so on, and which are injurious to the environment.

In this review, "toxinology" means the science of natural substances produced by or accumulated in living organisms, and the properties and the biological significance of these substances for the organisms involved. "Toxinology" (a word used in Swedish, though not often in English) is the author's homemade definition of a limited part of the more extensive concept of toxicology.

Venoms are often very complicated cocktails consisting of numerous different ingredients. Everything from simple organic substances to complicated enzymes may be included. One or more of these exert the toxic effects and constitute the toxins of the venom. Other substances, for example, hyaluronidase and anticoagulants, facilitate the penetration and spreading of the toxin away from the site of introduction. A number of other agents create a milieu which supports the toxic activity. Venoms vary in their chemical composition and in the physiological mechanisms they affect. Neurotoxic, cardiotoxic, hematotoxic and necrotic effects are common. In some animals, e.g., snakes, the venom contains enzymes which play an important role in the digestion of the prey.

2

The Study of Animal Venoms - Toxinology - Is an Important Science

Toxins of various kinds have proved to be indispensable tools for studying different physiological mechanisms. Because of the high specificity of certain toxins, they have made it possible to resolve discrete links, often at the molecular level, in highly complex biological mechanisms. Toxins have been used in various fields, such as neurobiology, pharmacology, muscle and blood circulation physiology. Several examples can be mentioned.

In the early sixties professor C. Y. Lee of National Taiwan University was able to isolate and purify the protein α-bungarotoxin from the Taiwanese krait (*Bungarus multicinctus*). This toxin specifically binds to the nicotine acetylcholine receptor (Figure 2), which mediates the signal from the nerve to the striated muscle and makes possible a controllable contraction. A unique chemical tool was now available, and investigations of the molecular and structural organization of the receptor began. This receptor is today one of the best characterized receptors in biology. The structure and amino acid sequence of its subunit has been described and its genes cloned. As a result, the disease myasthenia gravis, an autoimmune syndrome caused by antibodies to the receptor,

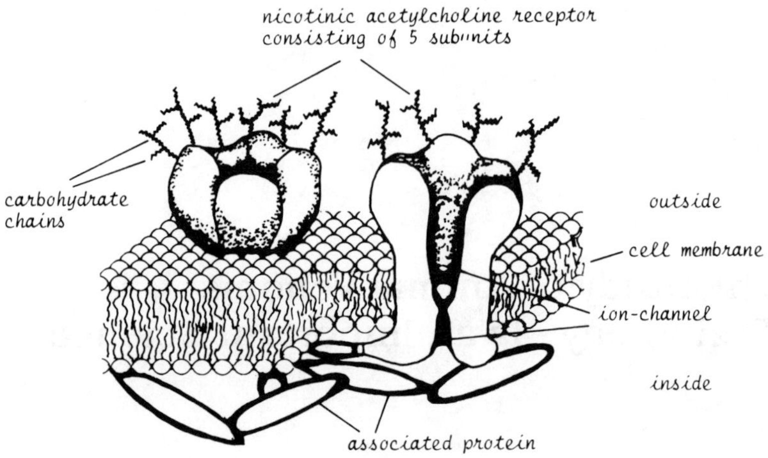

Figure 2. Postulated construction of the nicotinic acetylcholine receptor. After Lindstrom et al. 1983.

leading to the muscle weakness, can now be studied at the molecular level. Without α-bungarotoxin, a tool which nature itself has produced, this knowledge would not have been achieved.

With the use of other toxins, the mechanisms of the action potential and especially that of the ion currents have been tackled. Toxins which have contributed to such information have been extracted from the puffer fish (tetrodotoxin), dart-poison frogs (batrachotoxin), scorpions and sea anemones. These toxins bind to different sites of the very complicated molecule which forms the Na^+ channel. In this way the Na^+ currents, so important for the action potential, can be analyzed and explained at the molecular level.

Certain snake venoms contain toxins which affect different steps in the complex sequence of events leading to blood coagulation. Some toxins activate and others prevent this process. Such toxins already are and are likely to be of increased importance as experimental tools. One "toxin" has even been successfully used in the treatment of patients having a disposition to thrombotic disorders. This is ancrod, a thrombin-like enzyme purified from the venom of the Malayan pit viper (*Agkistrodon rhodostoma*). In Europe it is commercially available under the name Arvin. Like thrombin, it acts as a coagulant in vitro by catalyzing the transformation of fibrinogen into fibrin. In contrast, it is an indirect anticoagulant in vivo by slowly converting the fibrinogen into very small clots of blood, microembuli, which can thus be eliminated from the circulation. The blood remains incoagulable until fibrinogen levels are renewed.

Long before the scientist Paul Müller in the forties had synthesized and put the insecticide DDT on the market, an efficient way was known in Japan to make use of naturally produced toxins to fight flies. The toxin which has been used for centuries is nereistoxin, produced by a marine annelid worm—*Lumbriconereis heteropoda* (Figure 3). In Japan, dried worms are used as means of combating flies. In contrast to DDT, a persistent toxic compound of which all living organisms today bear traces, nereistoxin is in the course of time rendered harmless through natural degradation processes. This is true of all toxins produced by biological organisms.

Several substances with antibacterial or antiviral properties have been found in the lower marine animals. One interesting and important example of this is cephalosporin C, and antibiotic agent which today has a broad pharmaceutical application (Figure 3). This substance was discovered quite accidentally. It has long been known that the seawater near a sewage outlet from a town on the coast of Sardinia underwent a dramatic and rapid purification process. On closer examination it turned out that the change could be ascribed to the presence of a microsponge, Cephalosporium, which possessed the ability to inhibit the growth of bacteria. One of the active sponge components was termed cephalosporin. This substance is a potent antibiotic, able to inhibit the vast majority of bacteria, similiar to penicillin, although cephalosporin is insensitive to the enzyme penicillinase, which slowly inactivates penicillin.

Cytarabine (cytosine arabinoside) is another interesting substance (Figure 3), which can be extracted from certain sponges. This compound inhibits DNA synthesis and is now used as an antitumour drug.

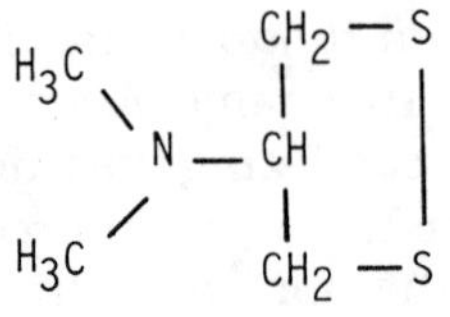

Nereistoxin

From the annelid worm, Lumbriconereis heteropoda

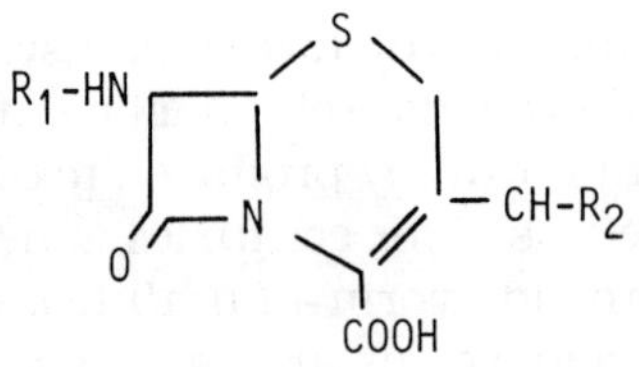

Cephalosporin

From the microsponge, Cephalosporium

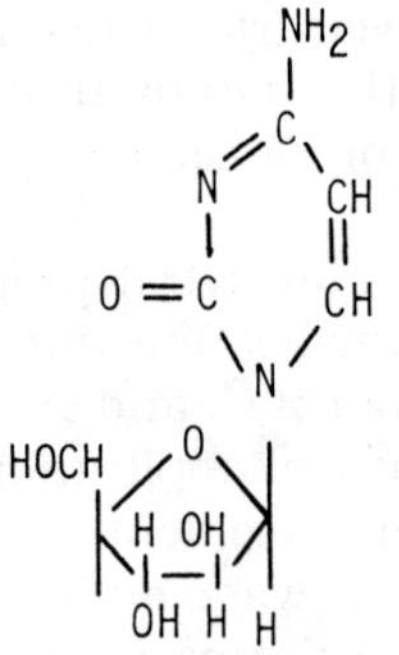

Cytarabine

From the West Indian sponge, Cryptotethya crypta

Figure 3. Interesting substances of some marine organisms.

Sponges and corals offer a rich selection of substances showing antibiotic properties. In the sea there is an enormous pharmacy at our disposal, and its exploration has just started. This is one of several reasons to protect the existing marine organisms and cease using the oceans as garbage dumps for toxic substances injurious to the environment.

During evolution, several animal species have developed resistance to the poisons of other animals or plants. The serum of the wood rat (*Neotoma micropus*) effectively neutralizes western diamond rattlesnake (*Crotalus atrox*) venom. The wood rat does not show any symptoms until the dose of venom is about 140 times greater than the amount lethal for an ordinary laboratory mouse. Certain invertebrate predators, in particular some gastropods, can consume highly toxic sponges, corals, and sea-cucumbers with no apparent discomfort. There are several other examples of an inborn resistance to animal poisons, but the mechanisms behind this phenomenon are largely unknown. Increased knowledge may contribute to the understanding of detoxication processes in our own body.

Many animal toxins have been identified today. In several cases we know their chemistry and mechanisms of action. However, modern toxinology is a young field of research. Many toxins, especially those of the marine organisms and of the insects, remain to be detected and their function determined. This research will yield a good profit in that we will gain an increased insight into the behavior and physiology of the animals. It will also contribute to a better understanding of our own physiological processes.

3

Experimental Pitfalls

The use of toxins as experimental tools has several pitfalls. One is that a toxin is usually present in a venom, that includes several biologically active components with individual mechanisms of action. It seems as if evolution has favored the development of pharmacologically more and more complex venoms. This trend is especially evident in the case of snake venoms. A complex venom presents problems when the toxin must be isolated from the other chemical substances. However, to purify a specific snake venom component free from, e.g., phospholipase A 2, is a chemically complicated procedure, requiring several chromatography steps. A contaminated toxin is a poor experimental tool.

Extreme specificity of a toxin is not always an advantage for the venomous animal. Therefore, in a venom several toxins may have evolved, each of which is characterized by its ability to affect various physiological mechanisms under different conditions. Apparently contradictory information in the literature may in certain cases be attributed to varying toxin concentrations, times of exposure or other factors in the different studies. An example of this is melittin, the dominating toxin in bee venom. Melittin appears to exert its action upon at least three different mechanisms depending on its concentration and time of exposure. At low concentrations, acting over a long period of time, it acts as a Ca^{2+} ionophore. At an intermediate concentration, for a short period, it activates phospholipases. At high concentrations it acts as a typical detergent, causing dissolution of the cell membranes. In all situations, the final result is irreversible damage of the exposed cell, but this takes place through activation of distinct initial processes.

4

LD50 Is a Common Measure of the "Killing Power" of a Venom

LD50 is a term used to describe the dose of toxin, which is lethal for 50% of the test animals over an observation period of 24 hours. LD50 is expressed in relation to some suitable unit of weight (often kg) of the test animal. The LD50 of a venom indicates its "killing power." A foreign chemical is usually considered to be a highly toxic substance if the LD50 is 1 mg/kg body weight or lower. A less common measure of toxicity is MLD (minimal lethal dose), which gives the smallest dose required to put an experimental animal to death. However, it is important to define the test species. Mouse, rat and rabbit can show very different sensitivities to the same toxin or venom. In addition, some laboratory animals are resistant to certain venoms, for example, cats and frogs toward cobra venoms. In this case, the mechanism of resistance is likely to be found at the site of the action of the toxin. Thus, the neuromuscular coupling of muscle preparations from the cat is insensitive to the otherwise so potent cobra neurotoxins.

With respect to toxicity tests, it is also necessary to describe the route of administration of the venom; intravenously (i.v.), intraperitoneally (i.p.), subcutaneously (s.c.), or in some cases intracisternally (intraventricularly) (i.c.). S.c. injection may often result in an impaired uptake and a higher LD50 for the toxin than when given i.v. Cobra venom (*Naja naja*) was reported to have the following LD50 values (mg/kg mouse): 0.13 (i.v.), 0.17 (i.p.) and 0.29 (s.c.).

When the poisonous dose exceeds the LD50, the number of experimental animals killed rapidly increases. A double dose usually kills all animals, whereas half the LD50 dose is without any effect. The toxic dose and the exposure time are often related in a way which is characteristic of the venom. An increased dose will reduce the length of time required to kill the test animal, although there is no simple linear relationship. Curves which illustrate relations of this kind may provide important information for an introductory description of an unknown venom or toxin.

5

Sponges (Porifera)

Sponges, the Porifera, are primitive and invariably sessile aquatic organisms. About 5,000 species have been described, and most of them are of marine origin. The majority of sponges are believed to produce toxic compounds that kill other organisms and help them to survive in an extremely competetive environment. Their skeleton is constructed of unappetizing structures, the so-called spicules, which consist of calcareous and siliceous materials. This contributes to their limited popularity for hostile animals.

From a utility point of view, these organisms are not only valuable as materials for bath sponges. A large number of interesting bioactive chemical substances have been isolated from sponges. These include antibacterial, antitumor, antiiflammatory, muscle relaxant, and hypotensive agents. Some of them have already found medical application, and there are reasons to believe that several other compounds of medical importance remain to be found in sponges.

The genus *Tedania* has attracted special interest. From one species, *Tedania digitata*, a nucleotide, 9-β-D-ribofuranosyl-1-methylisoguanine, has been isolated (Figure 4). This substance has been found to have potent muscle relaxant and cardiovascular

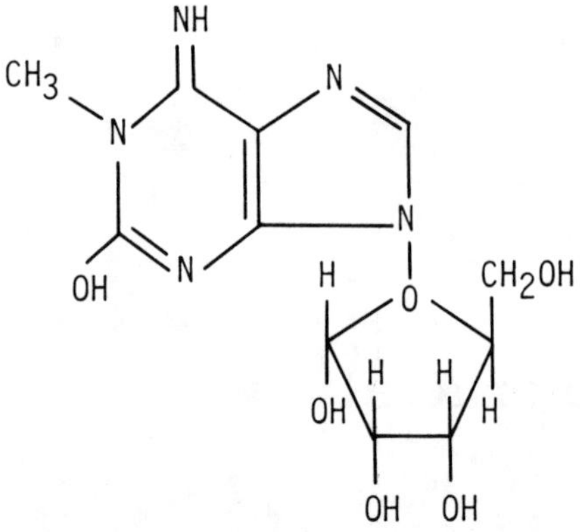

9-β-D-ribofuranosyl-1-methylguanine

Figure 4. A bioactive substance from the Sponge genus, Tedania.

effects. A closely related species, *Tedania ignis*, which is known to cause inflammation of the skin when touched by humans, has, strangely enough, a completely different composition of toxins. From *T. ignis* two neurotoxic fractions have been isolated; one of them inhibits the evoked release of acetylcholine in the neuromuscular nicotinergic coupling, and the other fraction induces the spontaneous release of acetylcholine. Because of the very polar properties of the fractions, it is likely, that they exert their effects on the outer surface of the cell membrane. Since the effects are also reversible, these active agents are of great interest for future studies of the physiology of neuromuscular transmission. This interest will increase once the bioactive components have been isolated.

6

Corals and Sea Anemones (Anthozoa)

Corals and sea anemones belong to the same class, comprising a large group of animals consisting of about 6,500 sedentary species. Among corals, colony-forming species dominate, whereas the sea anemones are solitary animals.

Corals

Like sponges, most corals contain several compounds of medical interest. Among others the prostaglandins can be mentioned. They exert a broad spectrum of biological effects, are very useful medically, and are expensive to synthesize. One species belonging to the horny corals, *Plexaura homoalla*, curiously makes such amounts of prostaglandins that the possibility of culturing these corals on a large scale for production of pharmaceutical preparations has been considered. Most coral species are probably also equipped with toxins for defense. Of 10 coral species examined during an 1981 investigation of invertebrates on the Great Barrier Reef outside Australia, 9 species contained substances having pronounced

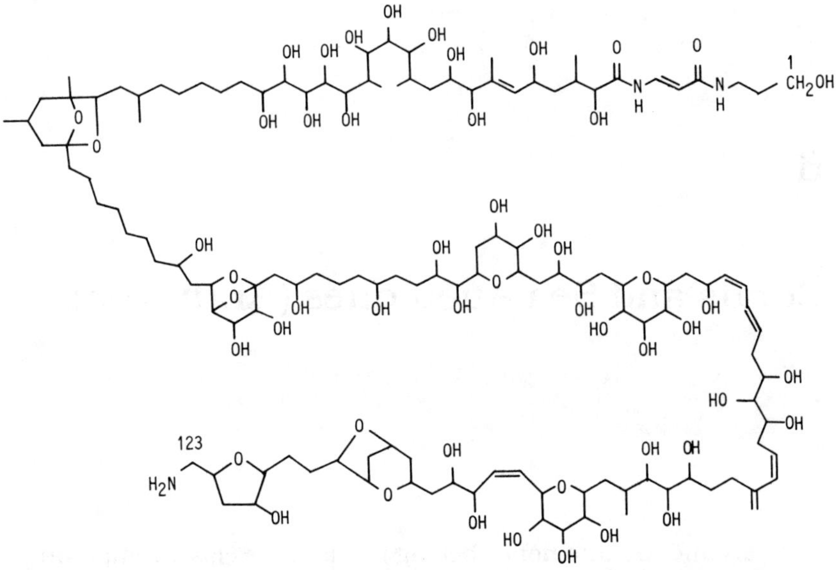

Palytoxin

Figure 5. Palytoxin is the most potent animal toxin known.

toxic effects on fishes. Most corals also appear to lack fish predators, although certain toxin-resistant gastropods prey on corals.

From various *Palythoa* species, which belong to the soft corals and inhabit the Caribbean and Pacific oceans, one of the most potent toxins known to toxinologists has been isolated. This substance is called palytoxin (PTX). Its true origin is unknown and also its pathway of biosynthesis. It is possible that it is synthesized by an unicellular organism and subsequently concentrated in *Palythoa* corals. Its LD_{50} in mice is 0.15 μg/kg at intravenous and 0.4 μg/kg at intraperitoneal injection. In comparison it can be mentioned that PTX is about 20 to 25 times more toxic than the extremely potent toxins tetrodotoxin (TTX) of puffer fish and batrachotoxin (BTX) of dart-poison frogs. Swimming in tidal pools inhabited by *Palythoa* can be dangerous. Contact with the coral slime in an open wound may cause a feeling of sickness, muscle cramps, difficulties in breathing, and could be fatal. Two scientists, Moore and Bartolini, were the first to elucidate the chemical structure of PTX (Figure 5). It has a molecular weight of 2,670, and is a long, partially unsaturated, aliphatic chain with interspaced cyclic ether groups. It is one of the most remarkable macromolecules in nature and is different from any other known molecule. PTX from different *Palythoa* species only show subtle differences in chemical structure. Shown to be an extremely effective depolarizing agent for all excitable cells, it is presumed to act as in ionophore, causing small pores in the cell membrane which results in a massive cellular inflow of Na^+ and outflow of K^+, and possibly also of other ions. It is believed that PTX transforms the sodium-potassium pump enzyme, Na^+, K^+ ATPase, to an ionophore. In support of this, ouabain, a plant glycoside which inhibits this enzyme, appears to partially counteract the PTX induced ion transport changes over the cell membrane.

Sea Anemones

From the toxinological point of view, the sea anemones are a very fascinating group of animals. They are equipped with highly specialized venom-containing stinging cells (nematocytes), which are triggered in order to catch prey or for defense (Figure 6). Stinging cells are common to all cnidarians, which include the Anthozoa, Scyphozoa and Hydrozoa classes. Their primary role is to introduce paralyzing and fatal toxins into the bodies of prey or potential enemies. The stinging cell is a capsular cyst, which contains a long slender tubule equipped with various accessory structures like stylets, barbs and lamellae. Upon triggering, the tubule is ejected from the cyst by evagination, during which the inside is turned out. When opposed to a target, the stylets punch a hole into the prey's skin and the venom is released.

From the stinging cells, tentacles, or other parts of sea anemones various peptides have been isolated. These molecules show neurotoxic, cardiotoxic, cytolytic and protease inhibitory properties. In particular, the neurotoxic substances have turned out to be interesting experimental tools in neurophysiological studies.

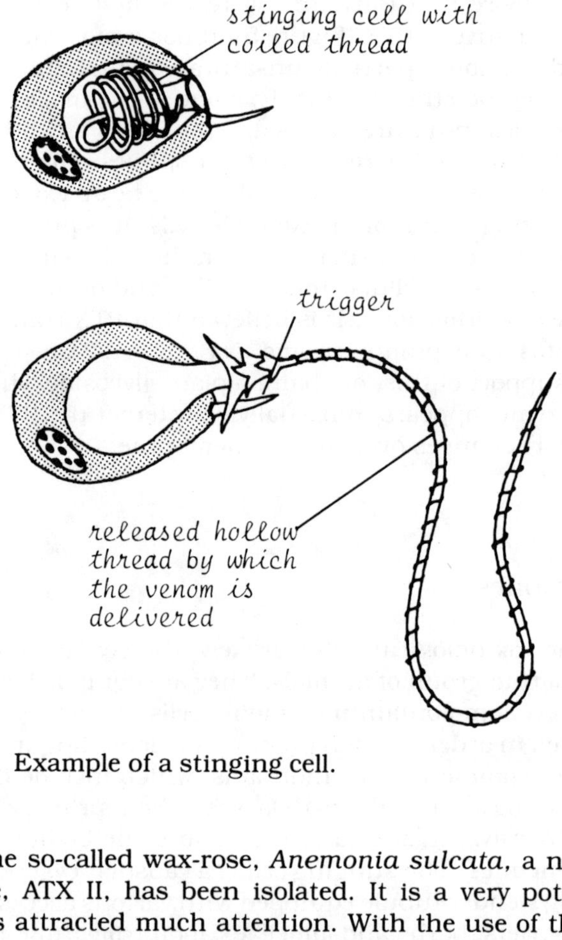

Figure 6. Example of a stinging cell.

From the so-called wax-rose, *Anemonia sulcata*, a neurotoxic substance, ATX II, has been isolated. It is a very potent toxin which has attracted much attention. With the use of this toxin, the wax-rose can paralyze prey such as fishes or various invertebrates. This sea anemone is a common cause of dermatitis among swimmers and divers in areas in which it is endemic. According to a somewhat older report, the wax-rose has even caused the deaths of some bath sponge divers in the Mediterranean Sea. However, the possibility of these deaths being due to secondary infections rather than to the venom cannot be excluded. The wax-rose inhabits the eastern Atlantic, extending from Norway and Scot-

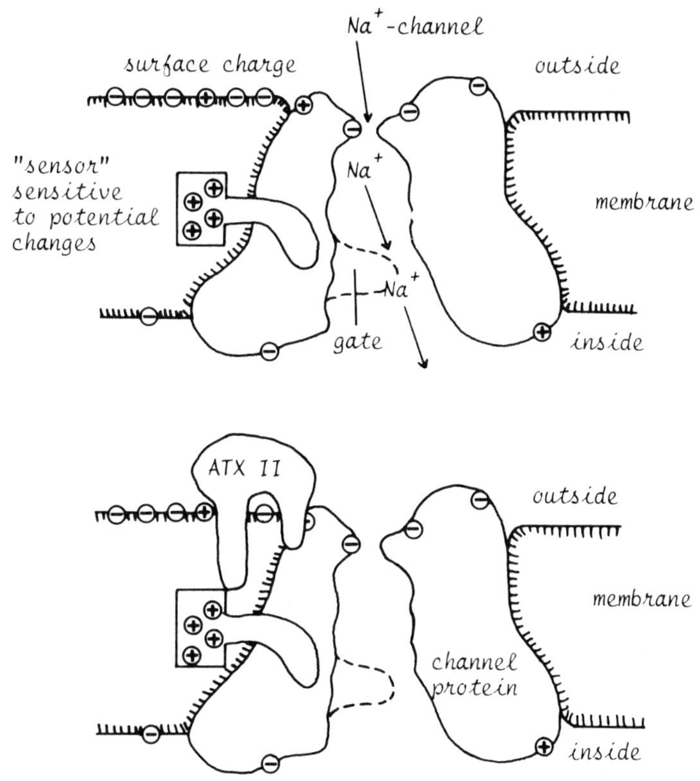

Figure 7. Proposed mechanism of action of a sea anemone toxin, ATX II.

land to the Canaries, and the Mediterranean Sea. It is a marine animal which is comparatively easy to keep in an aquarium, a further advantage for studying its toxic components. ATX II is a basic polypeptide, consisting of 47 amino acids and having a molecular weight of 4,770. It is characterized by a very high affinity to the voltage-dependent Na^+ channel and alters the inactivation of the Na^+ permeability, making it slow and incomplete (Figure 7), thus increasing the duration of the action potential. ATX II is included in the important arsenal of various neurotoxins, which have had a great impact in elucidating the voltage sensitive Na^+ channel. During recent years, a fluorescent derivative of ATX II has been synthesized with biological activity preserved. Assess to this molecule means further experimental advantages.

Another interesting toxin, equinatoxin, has been isolated from the beautiful, red colored, purple rose (also called horse-actinoid), *Actinia equina*. This species is found in the eastern Atlantic, extending from the Arctic Ocean to the Gulf of Guinea, Mediterrean Sea, and the Black Sea. Because of its display of color and ease of handling, the purple rose is a popular aquarium animal. The toxin is a basic, heat resistant, protein of 147 amino acids, with a molecular weight of about 20,000. It is very potent and has cardiotoxic effects. The LD_{50} for rats has been estimated to be 33 μg/kg for intravenous injection (i.v.), causing a rapid fall in blood pressure, bradycardia, followed by respiratory arrest. When given at a concentration of 100 μg/kg, i.v., pulmonary edema and hemorrhage occur. Because of this, equinatoxin has been used to study fluid regulation in the isolated lung. Under these conditions, it is extremely potent and at a concentration as low as 4 nM it produces marked changes in the fluid balance of the lung. In some unknown way, the toxin appears to break up the cell membranes of the pulmonary blood vessels allowing fluid movement into the alveolar spaces, which results in edema. The effects of the toxin can partially be inhibited by a special, and for the nervous system characteristic lipid, sfingomyelin. Equinatoxin, available in an easily cultured aquarium animal, may turn out to be a very useful tool in future studies of fluid regulation in the lung.

Toxins similar to equinatoxin have been found in several other sea anemones, for instance, *Stoichactis helianthus*, *Condylactis gigantea*, *Epiactis prolifera*, *Tealia lofotensis* and *Anthopleura xanthogrammica*. The toxins may vary somewhat in amino acid composition but the molecular weight is always in the range of 15,000-20,000. Like equinatoxin, they all have cytolytic effects which are inhibited by sfingomyelin. The mechanisms of action are unknown. It has been suggested that they cause transmembrane channels.

The sea anemone (Sea pink), *Metridium senile*, resides in the northern Atlantic and in corresponding climate zones in the Pa-

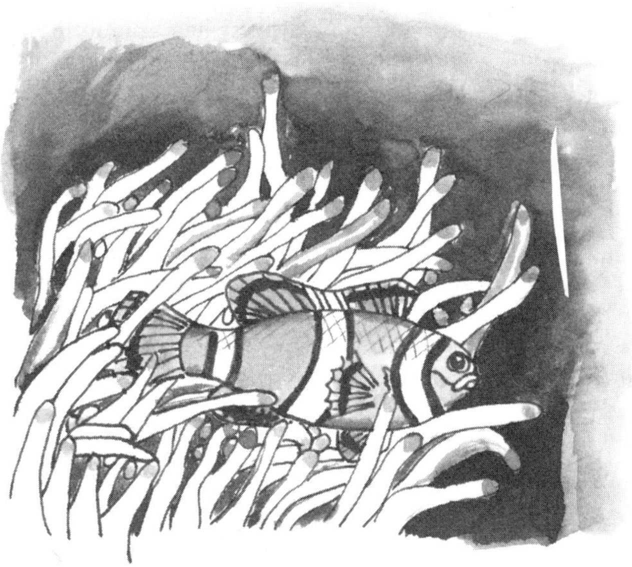

Figure 8. Some fishes, for instance clownfishes, get protection among the venomous tentacles of sea anemones.

cific Ocean. A cytolytic protein has also been isolated from this species. Its molecular weight differs from the other cytolysins but its mechanism of action may be similar. It is called metridiolysin. Electron microscopy studies have shown that it has the capacity to produce in cell membranes very small and regular 33 nm ring-like structures, which resemble those caused by certain bacterial lysins, of which streptolysin is the prototype.

The most venomous of sea anemones is probably *Rhodactis howesii*, which is found in the Indian and Pacific oceans. It is cooked by the inhabitants of Polynesia and appreciated as food. Carelessness is connection with cooking has had fatal consequences for several consumers of this both edible and dangerous sea anemone.

In the complex sea anemone venoms there are, besides the toxic components, often protease inhibitors present. However, their biological role in this context is unknown. It has been suggested that the inhibitors protect the tissues of the sea anemone from degradation by the proteases of the prey. A more likely explanation is perhaps that the inhibitors prevent the toxic components—the ammunition—from destruction.

The biggest species among sea anemones are found in tropical seas. Several of them belong to the genus *Stoichactis* (*Stichodactyla*). Species of this genus can measure a distance of more than a meter between the tips of the extended tentacles. They have attracted special interest since they appear to act as body guards for clownfishes (Figure 8). The fishes do not travel far from their friends and in case of danger they will quickly seek protection among the tentacles of their host animals. These fishes are possibly protected by a substance in their skin which inhibits the discharge of the stinging cells. Alternatively, they have developed an immunity toward the toxins.

7

Hydroids (Hydrozoa) and Jellyfishes (Scyphozoa)

Hydroids show more variations in anatomy and way of living than the jellyfishes. These groups generally have both a polyp and a free-swimming medusa generation, of which the latter stands for reproduction. The jellyfishes, which often inhabit the surface water, usually have larger medusae than the hydroids, which are also represented by several species living in the bottom layers.

The Portuguese man-o-war (*Physalia physalia*) belongs to the hydroids, even if it is commonly mistaken for a jellyfish. It inhabits the tropical Atlantic but is often brought in large numbers by the Gulf Stream as far north as the Hebrides. Its peculiar appearance and poisonous properties have made it famous. The name is due to the characteristic iridescent balloon-like float, from which meter long tentacles extend down in the water. The tentacles are equipped with large venom containing stinging cells. There is no detailed information available about the toxic components, which seem to be similar to those of common jellyfishes and which cause cytolysis. The Portuguese man-o-war is the most dangerous of the hydroids, accounting for numerous severe cases of illness with occasional fatal outcome. Accidents often occur in seaside resorts,

Figure 9. Portuguese man-o-war: A little fish, Nomeus, spends its life among the venomous tentacles.

the areas where they are seasonally plentiful, for instance in Florida and Hawaii. Besides an intense stinging pain and generalized weakness, they can have respiratory and cardiovascular effects. However, one little fish, *Nomeus*, appreciates the existence of this dangerous hydroid. This fish has developed some kind of resistance to the stinging cells and spends almost its entire life among the venomous tentacles, which protect the fish from predators (Figure 9).

Among the jellyfishes, some species have been subjected to a more thorough examination. These are the sea wasp or box jellyfish (*Chironex fleckeri*), a second species also called the sea wasp (*Chiropsalmus quadrigatus*), and the sea nettle (*Chrysaora quinquecirrha*). They are distributed in temperate and tropical seas. The sea wasp (*Chironex*) is regarded as the most dangerous, and perhaps the most venomous of all sea creatures. Fortunately, it is only found in the waters of northern Australia, where it has been reported to be responsible for an average of one death per year since 1900. Contact with the sea wasp gives rise to similar, but often more severe manifestations, than those noted for the Portuguese man-o-war, whereas the sea nettle is less toxic.

The stinging cells of jellyfishes contain several bioactive components. The most active and lethal compound appears to be a protein with a molecular weight of about 20,000. Some investigations suggest that it is inserted in the cell membrane and causes the opening of transmembrane channels for monovalent cations. Similar, but not necessarily identical toxic proteins, have been found in other jellyfishes, for instance, in the well known sea blubber (*Cyanea capillata*), in the Portuguese man-o-war and also in some previously discussed sea anemones. Even the phylogenetically distant spider, the Black Widow (*Latrodectus mactans*), may be equipped with a related toxin, although the molecular weights are quite different.

8

Polychaetes (Polychaeta)

Polychaetes are elongated, round or flat worms with a distinct outer and inner segmentation. The anterior part of the gut, the pharynx, is eversible and serves for grasping food. It is called a proboscis and can attain a length of several centimeters. In certain species, the proboscis has been developed into a stinging mechanism. The proboscis tip is armed with four symmetrically arranged chitinous jaws equipped with teeth, and each jaw is associated with a venomous gland. With the use of the jaws these animals can inflict a painful and toxic bite.

One species, *Glycera convoluta*, which is found in the Mediterranean Sea, is particularly interesting. From extracts of the venom glands, a toxin, α-glycerotoxin, has been isolated and shown to be a globular protein with a molecular weight of about 300,000. This toxin triggers the quantal release of acetylcholine (ACh) from the nerve terminals. Its mechanisms of action are still unknown, but differ from those of other toxins affecting the cholinergic transmission. It has been compared to latrotoxin from the Black Widow spider, which depletes the nerve terminals of their transmitters. In contrast to latrotoxin, the effects of α-glycerotoxin are reversible and the molecular weights of the two toxins are also completely different. Latrotoxin is presumed to act as an ionophore for monovalent cations (compare with jellyfish toxins), causing depolarization and influx of Ca^{2+} into the cell, followed by transmitter release. In contrast, α-glycerotoxin does not affect the permeability of the ions and seems to be a unique tool for studying cholinergic transmission. Strangely enough, there is no α-glycerotoxin in a related polychaet, *Glycera dibranchiata*. On the other hand, this animal contains another potent toxin, which appears to be similar to the very spider toxin, latrotoxin.

9

Gastropods (Gastropoda)

From an evolutionary point of view the gastropods are a particularly successful group of animals being represented by about 80,000 species. This means that it is the dominating class among the 100,000 mollusc (Mollusca) species. Their success can probably in part be ascribed to ingenious defensive and offensive weapons in which various venoms have an important role, and which compensate for their limited ability to move, their snail's pace.

The gastropods have an asymmetrical body enclosed by a spirally coiled shell. Some species lack a shell. The assymetrical organization is also true for the nervous system and other interior organs. They have a distinct head, bearing eyes and one or two pairs of tentacles. There is a large, broad and fleshy foot. Characteristic for this class is also the radula, a rasping, ribbon-shaped muscular tongue, studded with minute chitinous teeth and useful for scraping off small particles of food. Their teeth constellations have been used for the taxonomic classification of gastropod species.

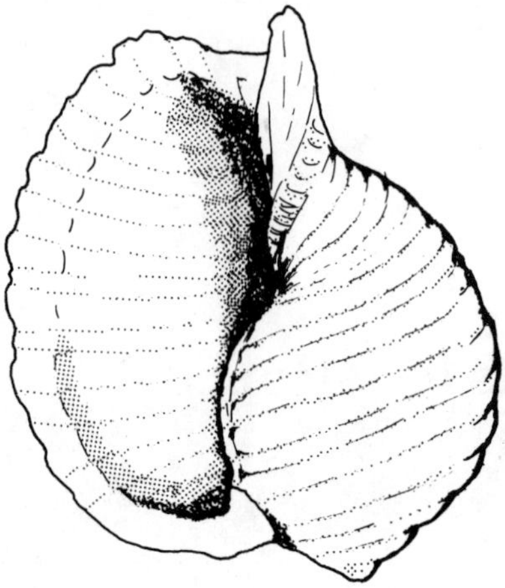

Figure 10. Tonna Galea, a "sulphuric acid spitting" mollusc of the Mediterranean Sea. Up to 25 cm, often smaller.

Tonnacea

The remarkable chemical equipment of gastropods was already investigated in 1853 by the well known German biologists, Troschel and Müller. When these scientists at a laboratory in Sicily were examining marine animals, two specimens of the Giant tun shell, *Tonna galea*, were brought to them by fishermen from the harbor (Figure 10). When they started to break one of the shells, the irritated mollusc stretched out its proboscis and ejected a clear fluid, saliva, which fell on the marble floor. To the surprise of the spectators, the saliva caused an intense corrosion of the marble. After a closer analysis of the fluid, these workers reported the existence of sulphuric acid spitting molluscs. This finding was confirmed 100 years later by professor Fänge, at a Department of Animal Physiology in Sweden. *Tonna galea*, and some other gastropods belonging to the family Tonnacea, possess this peculiar ability to secerete very acid solutions, contain-

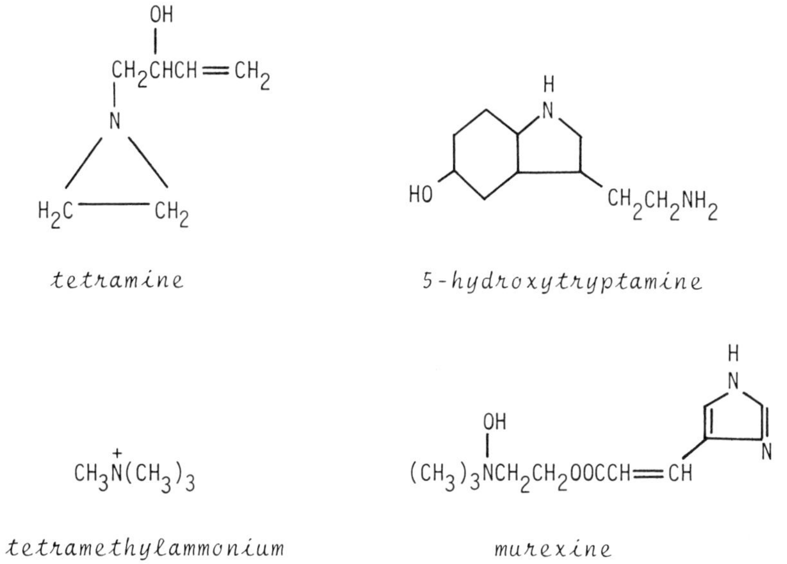

Figure 11. Some bioactive substances of gastropods.

ing about 3% sulphuric acid. The solution is squirted as a jet of clear fluid from an extended proboscis.

The mechanisms for the production of this sulphuric acid are still shrouded in mystery. Besides sulphuric acid, the fluid contains some toxic components. The acid probably softens the shell of the prey, for instance, the calcareous plates which enclose prey, such as certain urchins, starfishes and smaller crustaceans. In this way, the introduction of other toxins, paralyzing the prey, is facilitated. The gastropod also uses its venomous secretion for defense. In the saliva of other species, which usually are classified among Tonnacea, i.e., in the genus *Argobuccinum*, tetramine and tetramethylammonium have been found (Figure 11). These substances have curare-like effects and inhibit the nicotinic neuromuscular transmission, thereby paralyzing the prey.

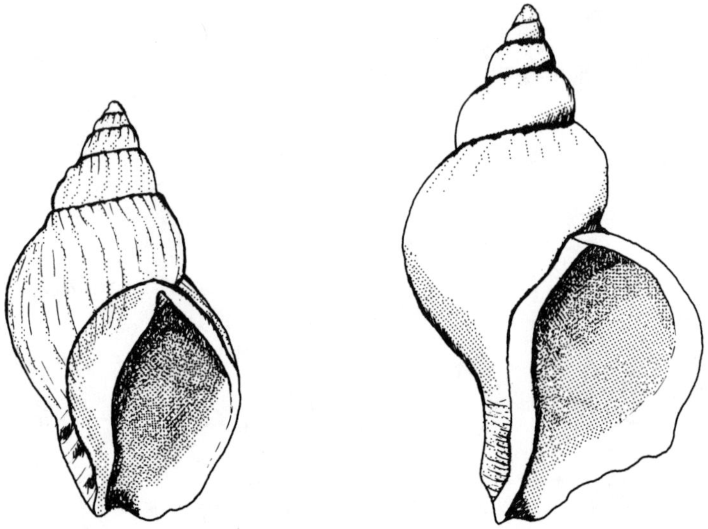

Figure 12. The poisonous red whelk (18 cm) to the right can be mistaken for the edible common whelk (10 cm) to the left.

Buccinidae

This family consists of a large number of carnivorous species, often armed with venom secreting glands. One example is the red whelk (*Neptunea antiqua*), which occurs in the waters of the northern Atlantic. This gastropod contains previously mentioned, tetramine and tetramethylammonium. On the other hand, the common whelk (*Buccinum undatum*), which is found in the same waters as the red whelk and which it somewhat resembles, is edible and delivered to fish dealers and restaurants in Europe (Figure 12). It has happened that by mistake red whelks have slipped into such deliveries, and, as a result some cases of poisoning have been reported. The symptoms are severe headache, a feeling of sickness or intoxication, a reeling gait, impaired vision, nausea and vomiting, and vertigo.

Another Buccinidae species, the Japanese ivory shell, has also given rise to cases of poisoning when eaten. The poisoning is likely to be due to the presence of the extremely potent toxin, tetrodotoxin (TTX) and a metabolite of TTX. It is most likely that TTX in the gastropod has an exogenous origin and is acquired through the food it eats.

Purple Whelks (Muricidae)

This is a family of carnivorous marine gastropods containing numerous species. The shells are often covered with spines and these animals are also interesting from the toxinological point of view. Several bioactive compounds have been identified in the secretion of their glands. Among these, murexine (urochanylchohline) and 5-hydroxytryptamine (5-HT) may be mentioned (Figure 11). Murexine shows nicotinic neuromuscular blocking activity.

Cone Snails (Conidae)

Cone snails represent the most fantastic and indigenous chemical "war machines," which nature has developed. There are about 500 known species and all produce complicated venoms, consisting of a whole arsenal of toxins (conotoxins) to immobilize their prey. There are three major feeding types of cone snails; worm eaters, mollusc eaters and fish eaters. These animals are called cone snails because of the characteristic cone-like shape of the shells, ranging in length up to about 25 centimeters (Figure 13). Cone snails are widely distributed in tropical and subtropical waters. They are often found in shallow waters, under rocks, along reefs, or crawling along the sand, and are among the major predators in tropical reef communities.

Figure 13. Conus litteratus is found in Australian waters. Up to 12 cm in length.

The ornate and attractive patterns of cone shells have made them much coveted by collectors and they often command a considerable price on the market. One dream snail is the very rare and famous *Conus gloriamaris* ("Glory of the Sea"), of which so far only 25 specimens have been found. Another one is *Conus cedonulli* (the matchless cone). A specimen of the latter was offered for sale at the Lyonet sale of 1796, along with Vermeer's painting, *Woman in Blue Reading a Letter.* This masterpiece was sold for 43 guilders, and the matchless cone, only 5 cm in length, brought 273 guilders.

The cone snail localizes its prey with a specialized chemical receptor, the osphradium, situated in the mantle cavity. This organ registers chemical changes in the water flowing into the mantle cavity. The venom apparatus consists of a muscular venom bulb, a long coiled venom duct and the radular teeth, which are housed in the radular sheath (Figure 14). Before stinging the radular teeth are passed into the pharynx and then into the pro-

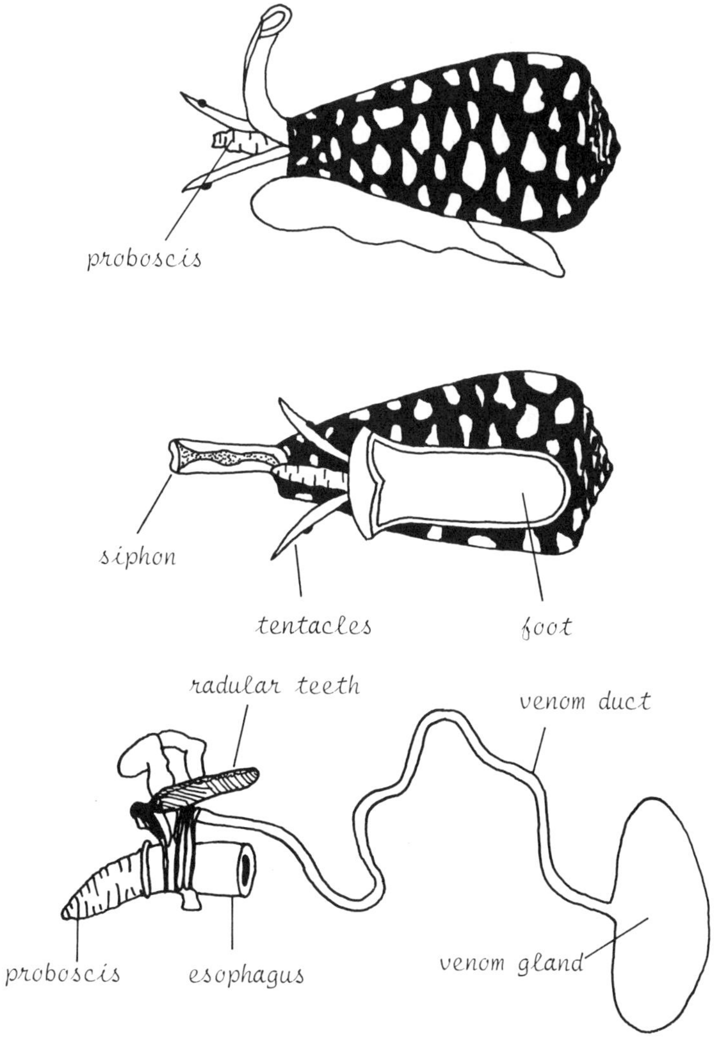

Figure 14. The cone snail's external anatomy. After Halstead, 1959.

boscis. The venom is secreted by pressure via the venom duct into the radula and thus into the lumen of the hollow radular teeth. The hunting snail is buried under the sand. When a prey is detected, the long fleshy proboscis moves slowly above the substrate. At the right moment the proboscis is rapidly extended and a venom containing tooth is thrust into the flesh of the victim. The tooth has a harpoon-like construction and cannot be dislodged by the prey. The teeth are disposable. There are always reserve teeth in the radular sheath waiting for new victims.

The venom apparatus is normally used for hunting, but may also be used for defense if the snail is threatened. The victim, for instance a fish, is immobilized within a few seconds after it has been hit and before it can manage to elude the slowly moving snail (Figure 15). The fish can then be swallowed with the snail's very stretchable and distensible proboscis. The elastic properties of the stomach makes it possible for the snail to digest a fish not much smaller than the snail itself. The venom is extremely potent and a sting may have serious consequences even to humans. In severe cases, it may cause paralysis, followed by coma and cardiac failure. The most dangerous of these snails is the *Conus geographus*, which lives in the Indian Ocean and is responsible for at least 20 recorded human deaths. These animals are sometimes referred to as "the beautiful but deadly cones."

Due to the intense research by physiologists and biochemists carried out in recent years, our knowledge of the properties of the unique *Conus* toxins is rapidly increasing. The venom is complex and consists of four classes of very specific toxins, which attack different strategic targets in the neuromuscular system of the victim and have paralytic effects (Table 1). With one exception, the toxins are relatively small basic peptides (13-29 amino acids), which are stabilized by one to three disulfide bridges. Besides the paralytic toxins, there are two peptides, one termed the "sleeper" peptide and the other conopressin. The "sleeper" peptide does not affect fish but induces a one to two day long sleep in mice. Conopressin is similar to vasopressin, which affects blood circulation in mammals by acting on the arterial smooth muscle. The reason for the presence of conopressin in the venom could be to redirect the venom containing blood of the prey, so that it is pri-

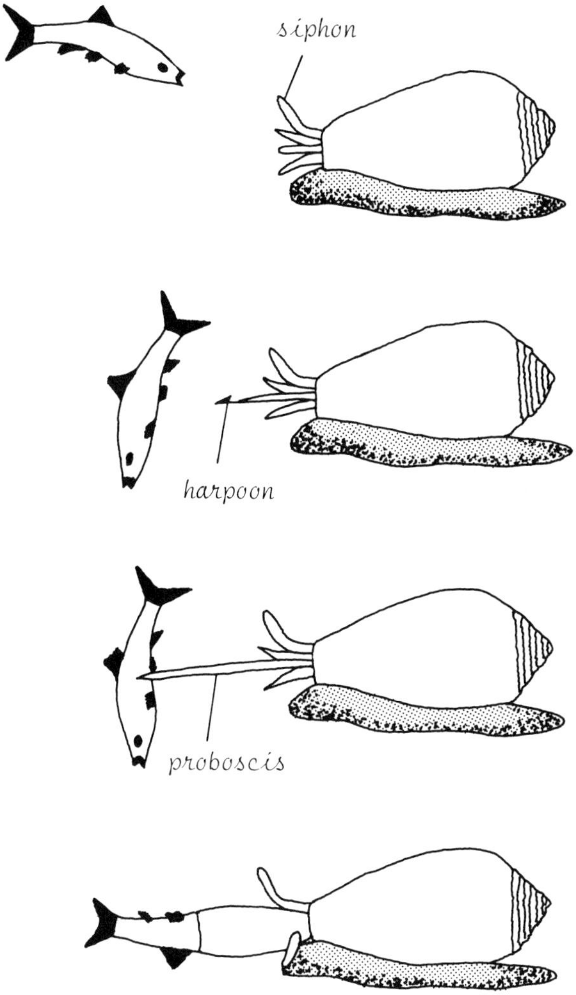

Figure 15. The cone snail catches its prey. After Fänge, 1985.

marily transported to the skeletal muscles. Another nonparalytic component is a 100,000 molecular weight protein, which provokes convulsions when injected in rats but apparently has no effect when administered to fish. The "sleeper" peptide,

Table 1. General properties of conotoxins. After Olivera et al., 1985.

	Venom source				
	C. geographus	C. magus	Symptoms in mice*	Physiological target	No. of amino acids
α-Conotoxins	+	+	Paralysis, death	Acetylcholine receptor	13–15
μ-Conotoxins	(+)	(+)	Paralysis, death	Muscle Na^+ channels	22
ω-Conotoxins	+	?	"Shaker"	V-sensitive presynaptic Ca^{2+} channels	25–29
K-Conotoxins	?	+	?	? K channels	~25
"Sleeper"	+	?	"Sleep"	Unknown	17
Conopressin	+	(+)	"Scratching"	? Smooth muscle	9
Convulsant	+	+	Convulsions, death	Unknown	~ 100

*Intracerebral injection

conopressin and the convulsant protein do not cause any visible physiological effects in fish when they are injected one by one. In spite of this, they are likely to have important accessory functions in the complex snail venom by facilitating the effects of the paralytic toxins, which will be presented next.

Two of the four classes of paralytic *Conus* toxins exert their actions postsynaptically. One of them is called α-conotoxin. It binds to the nicotinic acetylcholine (ACh) receptor in the neuromuscular coupling. The name-prefix (α) is based by analogy with snake toxins (α-bungarotoxin) showing similar affinity to this receptor. The other postsynaptic *Conus* toxin is termed μ-conotoxin. It preferentially blocks the sodium channel in the muscle but not in the nerve. Another type of paralytic toxins are the ω-conotoxins, of which there are different types. One of them is well characterized and acts by blocking the voltage activated calcium channels, which leads to inhibition of the release of ACh. Another toxin, K-conotoxin, has also been found, which appears to inhibit both pre- and postsynaptic K^+-channels. These paralytic toxins offer unique tools for future research in neuromuscular physiology and on the properties of the various ion channels in general. One of the ω-conotoxins has been commercially available for some years and has contributed to the characterization of the important Ca^{2+} channels.

The *Conus* venom shows remarkable properties. Firstly, the toxins are much smaller molecules than the toxic peptides or proteins found in other venomous animals, for instance, in snakes, scorpions, spiders and sea anemones. Secondly, the venom has an "over-kill" capacity by its content of several very potent toxins, acting on different vital sites of the neuromuscular system. The advantage of small toxic compounds could be that such molecules more easily penetrate into the tissue of the victim. Furthermore, small molecules can be present at a higher molar concentration, which is an advantage when the volume of the venom is small. The presence of several toxins, each killing, is probably due to the fact that the snail in relation to the victim, for instance a fish, is rather immobile. In order not to lose the prey, an almost instantaneous paralysis of the fish is required. From sting to immobilization of the victim it only takes some seconds, which is enough for the snail to secure the fish with the help of the extensible proboscis. By access to several toxins, which inhibit consecutive physiological events in the neuromuscular system of the fish, the chances increase that at least one vital process will be brought out of function, which is enough to paralyze the victim. Several different toxins also lead to synergistic toxic efforts. In these ways the snail can compensate for a small volume of venom.

Sea Slugs and Sea Hares (Opisthobranchia)

Opistobranchs form an interesting subclass of Gastropoda, consisting of about 3,000 species. Their shell is reduced or lacking and they show a great adaptive range of form, color, feeding and locomotion. These animals have developed defensive mechanisms, which compensate for the lack of a protective shell. The so-called nudibranchs entirely lack a shell and are the largest and most slug-like of the opistobranchs. In his book, *The Log From The Sea Of Cortez*, John Steinbeck writes about his friend the marine biologist

Edward Rickett: "With any new food or animal he looked, felt, smelled, and tasted. Once in a tide pool we were discussing the interesting fact that nudibranchs, although beautiful and brightly colored and tasty-looking and soft and unweaponed, are never eaten by other animals which should have found them irresistible. He reached under water and picked up a lovely orange-colored nudibranch and put it in his mouth. And instantly he made a horrible face and spat and retched, but he had found out why fishes let these living tidbits completely alone."

Nudibranchs, like many other opisthobranchs, use repellent skin secretions to protect themselves against predators. Some of these species have no capacity to produce their own venoms. They have developed a sophisticated technique, exploiting the venoms of other animals for own purposes, and using stinging cells (nematocytes) as weapons. The stinging cells are supplied by different coelenterates on which they feed. The Sea Slugs are obviously immune themselves against the toxic components of the stinging cells. When they have consumed a coelenterate, the stinging cells are not digested, but stored in a gland associated to the gut. The gland is connected to skin protuberances, to which the stinging cells are delivered. The cells next grow into mature stinging cells ready to be used against hungry enemies or for offensive purposes.

Because of its pelagic way of living, the blue species, *Glaucus atlanticus*, has attracted much attention in the literature. This Sea Slug lives in tropic seas and is a predatory animal, which utilizes the highly toxic stinging cells of Portuguese man o'war (*Physalia*) as effective weapons.

Other species (*Pleurobranchus, Berthella, Philline*) produce an acid secretion from skin glands, which they release when the animals are disturbed. The secretion contains, besides sulfuric acid, various organic compounds, of which little is known so far.

The Sea Hare (*Aplysia rosea*) releases a purple colored secretion when it is in danger. The color is due to the presence of chromoproteins, produced by a special gland. Even other toxins, probably of dietary origin (blue-green algae), may be secreted.

10

Octopods (Cephalopoda)

Cephalopods include the nautilus, squid, cuttlefish, and octopus. They are the largest and most complexly organized of the Mollusca, and probably also the most highly organized invertebrate animals. In several respects they have developed in parallel to vertebrates. There are about 750 living and more than 10,000 fossil species. They are free-swimming predators containing a large head, a complex brain, and conspicuous and well-developed eyes. The mouth is armed with horny jaws and surrounded by eight (Octopods) or ten tentacles equipped with several suckers or hooks. Rapid movements are produced by expelling water under high pressure through the siphon. For toxinologists, some Octopods are especially interesting.

Few marine animals have perhaps attracted more interest from the public at large than has the octopus. Despite all fables about these animals, they are in fact shy and reclusive organisms. With the exception of some Australian species, they hardly expose man to any danger. The best known of the octopuses is the common octopus, *Octopus vulgaris*, which is found in many parts of the world. Like other octopods, it seizes the prey with the tentacles. The jaws have a special construction to bite the victim with great force. Salivary glands, which discharge into the mouth cavity,

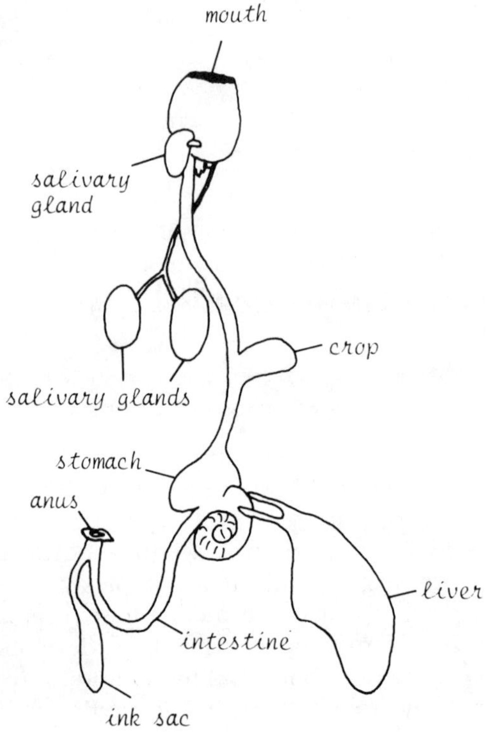

Figure 16. The secretory system of Octopus. After Halstead, 1959.

secrete a venom that immobilizes or kills the prey (Figure 16). Digestive enzymes are also injected into the prey, for example, a crab, which some hours later is expelled, apparently quite intact. However, only the shell is remaining. All internal organs have been digested and sucked out.

The posterior salivary glands of the octopus secrete a strongly acidic fluid, which contains a component toxic for crabs. Furthermore, there are several physiologically active molecules like tyramine, octopamine, dopamine, noradrenaline, serotonin (5-HT), histamine, taurine and acetylcholine (ACh). Some of these are likely to be normal constituents of the salivary glands and not necessarily venom components. Proteolytic enzymes, hyaluronidase, and a special glycoprotein (cephalotoxin) are also present, the exact composition varying among octopus species. One bioactive peptide, termed eledoisin, has been identified in the posterior salivary gland of *Eledone moschata*. It is pharmacologically very potent, causing marked vasodilation, hypotension and increase in the permeability of peripheral blood vessels. The role of this substance in *Eledone* is still unknown.

The crab-killing activity of the venom is not likely to be explained by the presence of the various amines. It is possible that the lethal toxin is the glycoprotein named cephalotoxin. It has been fractionated into five components, of which two are acidic glycoproteins. Their mechanisms of action are still unknown. Humans who have been bitten by *Octopus vulgaris* usually complain about pain but generally recover rather quickly. No deaths due to poisoning have been reported.

The Blue-ringed octopuses *Hapalochlaena (Octopus) maculosa*, and *H. lunulata*, are mainly found in the waters off the coast of Australia and are common in tidal pools. These seemingly sweet little things (9-20 cm long, weight about 90 gram) are beautiful to look at and it is tempting to pick them up and play with them. This is, however, not recommended. They have remarkably large posterior salivary glands containing a venom dangerous to man and some fatalities have been recorded. The active component is a neurotoxin called maculotoxin. It is known to inhibit activation of the Na^+-channel in nerves and appears to be identical to tetrodotoxin.

11

Echinoderms (Echinodermata)

Echinoderms represent a big phylum, which includes sea urchins, starfishes, sea cucumbers, and their relatives. It consists of about 7,000 marine species. They are mostly unaggressive, slow moving animals, which live on the bottom of the sea. They are widely spread in all seas and exist at varying depths. In quantity they form the dominating part of the fauna of the deep-sea bottoms. These animals show a bilateral symmetry as larvae and the characteristic radial symmetry as adults. In the adults the radii, which are nearly always five, diverge from the mouth. The radial symmetry is most conspicuos in starfishes. The surface consists of calcareous plates, that often bear external spines and other processes. Among echinoderms, sea urchins (Echinoidea), starfishes (Asteroidea) and sea cucumbers (Holothuroidea) are especially interesting from the toxinologic standpoint.

Sea Urchins (Echinoidea)

About 600 sea urchin species are known today. Like starfishes, the sea urchins have pedicellariae, or modified spines, which arise among the ordinary spines from the calcerous plates. The

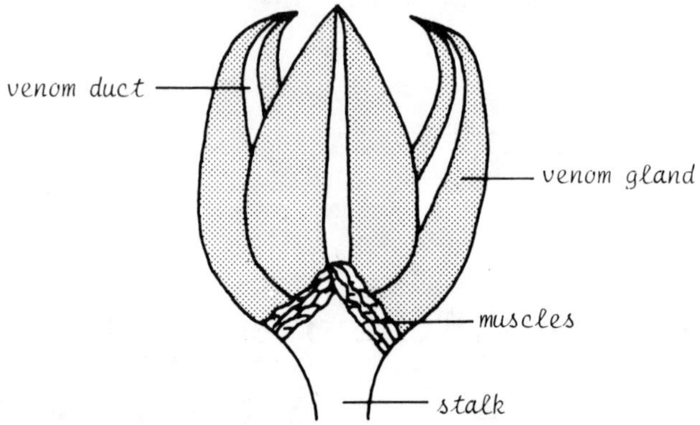

Figure 17. A pedicellaria from sea urchin.

pedicellariae are about 1-4 mm long and there are different types. They serve for self-defense and in some species are associated with venomous glands. The pedicellaria consists of a supporting stalk and a conical head, which is armed with a set of calcerous jaws forming a seizing organ, the movements of which are regulated by strong muscles (Figure 17). When the animal is disturbed, venom is pressed along ducts into the hollow tooth-like tips of the jaws. At the same time the victim is seized by the jaws, the venom is injected. Stings from some species are very toxic. A common and feared species is *Toxopneustes pileolus*. It inhabits the sea from East Africa all the way to Japan and the Fiji Islands and Japanese fishermen hold this sea urchin in great respect. They consider it responsible for several cases of severe poisoning in man. Its venom causes respiratory distress and dizziness. The toxic factor is a protein. Its chemistry and mechanisms of action remain to be elucidated.

Starfishes (Asteroidea)

Starfishes, of which there are about 2,000 described species, are to a greater extent than sea urchins active predatory animals

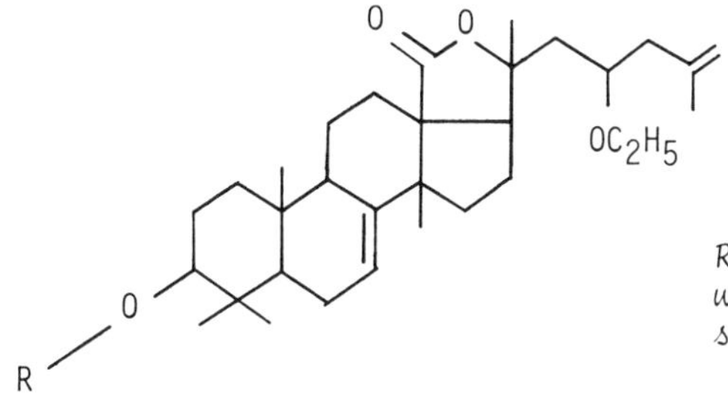

R = simple carbohydrates, which vary in different steroidglycosides

Steroidglycoside

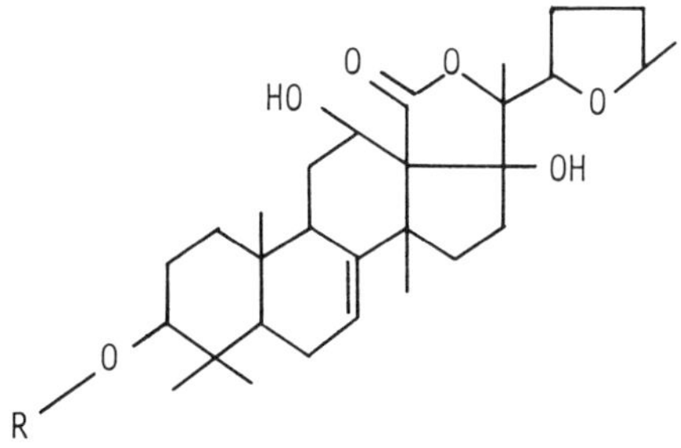

R = simple carbohydrates, which vary in different triterpenglycosides

Triterpenglycoside

Figure 18. Toxins of starfishes and sea urchins.

and are equipped with potent toxins. The active components have neurotoxic effects and produce an irreversible blockade of cholinergic neuromuscular transmission. These toxins are steroidal glycosides (astichoposide C, thelenotoside B, asterosaponin L, etc.), which affect the passive membrane permeability of ions and presumably also their active transport by inhibition of Na^+, K^+-ATPase activity (Figure 18). Like the plant glycoside ouabain, which these toxins chemically resemble, they cause depolarization of cell membranes.

Sea Cucumbers (Holothuroidea)

Sea cucumbers form another group of echinoderms of which there are about 2,000 known species. They use similar but more potent toxins than the starfishes. The toxins are various triterpene glycosides, possessing Na^+, K^+-ATPase inhibitory effects but also interfering with several other biological mechanisms. They have been reported to exert neurotoxic, cytotoxic, antifungal and antivirus effects. There are several different triterpene glycosides (holothurin A, B and C, cucumarioside G, etc.) (Figure 18). Some sea cucumbers also ingest coelenterate stinging cells, which are later used for their own defense.

The inhabitants of tropic islands use squashed venomous cucumbers to paralyze fishes. In spite of this, several species of the genus *Holothuria* and the genus *Stichopus* are regarded as gastronomic delicacies. Sea cucumbers are the basis of trepang, a dish which in China is regarded as a great delicacy. They also think there is a further advantage of this dish. It has the reputation of being sexually exciting, an aphrodisiac. China imports large quantities of sea cucumbers mainly from the West Indies and Japan.

The sea cucumbers are exposed to keen competition by other bottom living animals, not the least from the sea urchins. The sea cucumbers sporadically release their venoms in small doses to the environment and it is known that these venoms at very low concentrations inhibit the development of the fertilized eggs of the sea urchins. It has therefore been speculated that the sea cucumbers, with the use of their toxins, suppress the reproduction of the sea urchins.

12

Insects (Insecta)

Insects (Insecta) belong to the arthropods (Arthropoda), which are characterized by a segmented body and an external chitinous cuticle. In insects, the body is divided into three distinct regions, the head, thorax and abdomen. The head consists of six segments and there is a single pair of antennae. The thorax consists of three pairs of legs and usually two pairs of wings. The abdomen is typically divided into eleven segments.

Insects, with its one million species or more, represent by far an outstanding class of animals with regards to the number of species. In comparison it can be mentioned that there are about 21,000 species of vertebrates. As far as the physiology of insects is concerned, most remains to be investigated and described. This is to a high degree true for the insect venoms. Among insects, the venoms of members belonging to the order of Hymenoptra (bees, wasps, and ants) are the best known. However, with some exceptions, our knowledge mainly concerns the venoms of social bees and wasps, which cause pain and inflammation. In contrast, little is known about the venoms and their mechanisms in the solitary species, like the rest of the insects. Intensified research can be expected to be important for the development of new principles to control insect populations and to give us new pharmacological tools for studying basic physiological mechanisms.

Bees (Apidae)

The family Apidae includes both social and solitary bees and bumblebees. Our knowledge of bumblebee venom is limited. In contrast, the venom of the honeybee (*Apis mellifera*) is probably the most thoroughly characterized insect venom. Because of the economic importance of the honeybee, its reproduction, behavior and general physiology have also been intensively studied.

The honeybee colony is defended only by its sterile females. The venomous system is a modified part of the genital system having defense as its only function (Figure 19). In contrast, the majority of venomous insects are predators which feed on living prey, which usually comprise other insects. The stinging behavior of the honeybee is only stimulated in the vicinity of the colony. The venom is produced by secretory cells forming a venom gland, and is stored in a reservoir containing about 1 to 3 μl venom. The venom is released under pressure through a duct terminated by a sting equipped with a barb, which will anchor firmly in the skin of the victim. As a result the honeybee will die, since associated glands and organs are torn out of its body together with the sting and its attached venom gland. The gland continues its automatic pumping action and injects more venom into the victim. Unlike bees, the wasps and ants survive because their stings lack barbs and do not remain in the victim. Some special odors are supposed to arouse the aggressive behavior of the honeybees. The aggression is mostly directed towards competitors for the food. Rivals are very often other anthropod species. One injection corresponds to about 0.5 μl fluid and is painful but mostly harmless for larger animals. For other insects it is generally lethal.

A massive attack by several bees may through physiological shock have serious effects for humans. Still more dangerous is

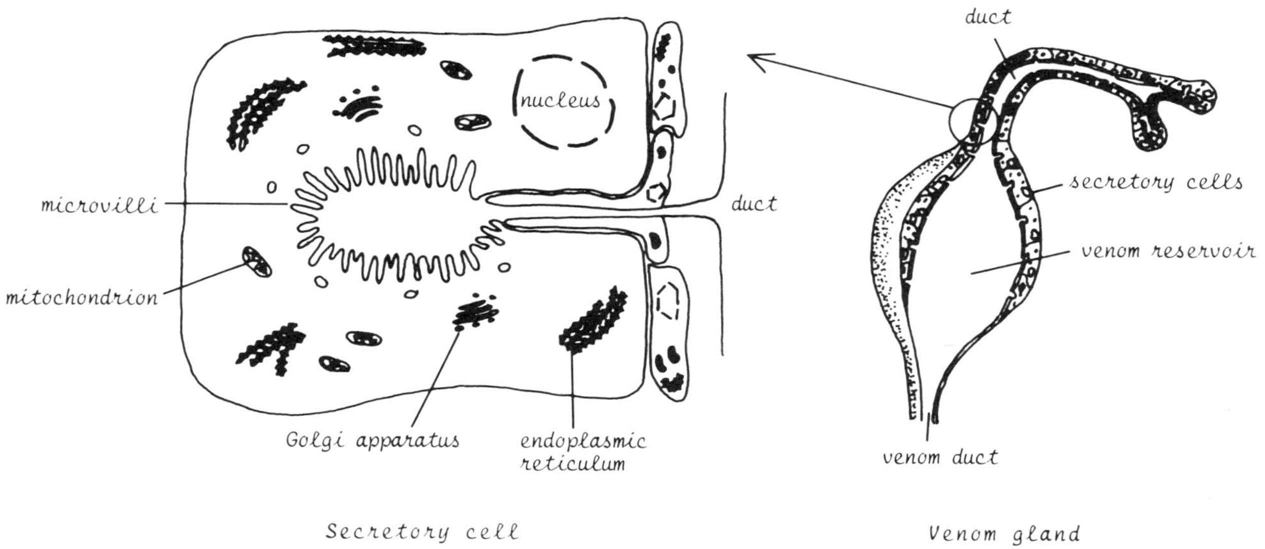

Figure 19. Venom gland and secretory cell from the honeybee. After Owen and Bridges, 1974.

sensitization to the venom, which may develop after repeated stings. As a result allergic reactions, including acute life-threatening anaphylaxis, may occur. For this reason the honeybee, which is so popular with many of us, is at the same time perhaps the most dangerous venomous animal in the world. In the United States, the stings of honeybees are supposed to account for more deaths in humans than the bites and stings of all other venomous creatures.

Ingenious electrical devices to collect bee venom in large amounts have made possible a detailed analysis of the venom's chemical composition. By exposing a beehive to weak electrical fields of high frequency, the bees can be stimulated to sting through a nylon mesh against a glass plate. After dying, the venom may be scraped off in quantities sufficient for chemical analysis.

The venom is very complicated. It consists of at least 40 different substances, with one injection (0.5 μl venom) containing about 50 μg dry substance. The major protein components are melittin (50% dry weight), phospholipases A (12%), "mast cell degranulating factor" (MCD) (3%), and hyaluronidase (3%). In addition, there are numerous bioamines, among them histamine (1%), dopamine (0.5%), and noradrenaline (0.5%). Furhermore, several volatile acetates are present, which are presumed to stimulate aggressive behavior in other bees.

Of the various components, melittin appears to cause the intense, localized pain, which we feel after a bee sting. Melittin is well characterized, being composed of 26 amino acids, which form an amphiphilic peptide (Figure 20). It has been suggested that in the victim, four melittin molecules come together and form a tetrameric pore in the phospholipid layer of the cell membrane. This construction requires the presence of a membrane potential. The pore acts as an ionophore, permitting the free passage of anions. This results in depolarization of the membrane and dissociation of the tetramer into inactive monomers, which are released from the cell membrane and diffuse away. How can this explain the sharp and stabbing pain typical of bee stings? According to one theory, the pain receptors in the skin, which are formed

H_2NGly-Ile-Gly-Ala-Val-Leu-Lys-Val-Leu-Thr-Thr-Gly-Leu-

-Pro-Ala-Leu-Ile-Ser-Trp-Ile-Lys-Arg-Lys-Arg-Gln-Gln$CONH_2$

Figure 20. Amino acid sequence of melittin from the honeybee (Apis mellifera).

by unmyelinated afferent nerve terminals, are depolarized due to the ionophoric properties of the tetramer. This elicits a repetitive electrical firing of the nerves mediating pain and is followed by an intense sensation of pain. When the nerve terminals depolarize, the tetramer dissociates into monomers, a membrane potential can again be established and the whole process be repeated. In time the monomers will diffuse away and the pain will fade out. Melittin is not particularly toxic to man. Therefore, it has been speculated that its pain-producing properties are mainly aimed at mammalian predators, who steal the honey which is so important for the survival of the bee colony. The biogenic amines in the venom may also contribute to the pain. It is possible that pain-producing substances are effective in developing avoidance reactions in vertebrates.

The swelling, which often follows a bee sting, is supposed to be due to the presence of phospholipases and hyaluronidase in the venom. Phospholipase A_2 is the major allergen in the venom and probably the factor responsible for the sometimes very severe consequences of bee stings for humans. In bee venom, like in so many other venoms, there is hyaluronidase. Its probable function is to facilitate spreading of the venom by disintegrating barriers to the diffusion of the other venom components. Complex intercellular mucopolysacharides constitute such barriers.

Another venom component, which has attracted considerable attention, is apamin, a small peptide consisting of 18 amino acids (Figure 21). This peptide can pass the blood-brain barrier and has neurotoxic but also more general physiological effects. Previously it was believed that it specifically affected synaptic functions in the central nervous system. However, today we know that apamin exerts effects on several cell types by blocking a particular class of

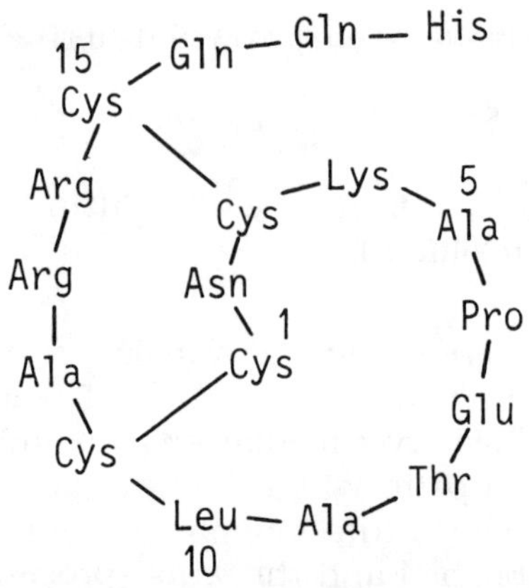

Figure 21. Amino acid sequence of apamin.

Ca^{2+}-dependent K^+-channels in the cell membrane. K^+-channels are very diverse, both in function and occurrence. Apamin and some other recently characterized toxins from scorpions and snakes are proving to be invaluable tools for determining what kinds of K^+-channels exist in various cells, and what the roles of the channels might be.

The mast cell degranulating factor (MCD) is a peptide which chemically resembles apamin. It is probably the major venom factor responsible for the massive release of histamine. At the same time, this peptide possesses the powerful anti-inflammatory properties associated with honeybee venom, and has therefore received considerable interest in medicine. Attempts to identify and isolate that part of MCD which exerts the anti-inflammatory effects have so far been unsuccessful.

Possibly with the exception of melittin, the venom components are likely to be aimed at other arthropods, such as predatory wasps and parasitic wax moths. The mechanisms of action in these animals are still unknown. The very small amounts of, for example, phospholipase, apamin, histamine and the other bioactive components which are released from the sting, can hardly have any physiological effects on larger animals.

Wasps (Vespidae)

There are about 15,000 described wasp species spread all over the world. The majority of them are solitary. Only venoms of a few solitary and some social species have been thoroughly examined. Wasps have barbless stings, and thus multiple stings can be inflicted without damage to the wasp.

Many of us fear wasps. The stings of some species can be extremely painful. Furthermore, social wasps are aggressive and whole wasp swarms can attack large mammals with fatal consequences for the victim. Allergic reactions to wasp venom may be as serious as those to bee venom, but are not as common, because fewer persons come in contact with wasps. Solitary wasps have developed an efficient method to paralyze their prey. This behavior was already known by the ancient Chinese. The Erh-Ya (Kuo Po, A.D. 276-324) describes a green "worm" which was paralyzed by a wasp. The prey, generally another insect, is stung in its ganglia or their vicinity. The ingredients of the venom immobilize but do not kill the victim, which is then carried to the nest and placed in some cell to serve as food for the growing larvae. The venom consists of neurotoxins, which vary among different species and are still relatively unknown. Acetylcholine and histamine are generally present in concentrations high enough to affect nervous functions. The venom also contains more complicated substances supposed to block the ion channels of muscle fibers and to be responsible for the paralysis. Research on the chemistry and mechanisms of action of these toxins is one among several fascinating future fields in toxinology.

Venoms of social wasps, in particular from some species of yellow jackets (*Vespula*), hornets (*Vespa*) and paper wasps (*Polistes*) are better known. Like honeybee venom, these wasp venoms contain biogenic amines, especially histamine, serotonin, dopamine, noradrenaline and polyamines. Of these, histamine dominates. The percentage composition of the amines varies in venoms of different species, which has even been the basis for some chemotaxonomic studies of wasps.

Besides the amines, there are several peptides and proteins in the venoms of social wasps. Among others, there also kinins, also

Sequence	Source
Arg-Pro-Pro-Gly-Phe-Ser-Pro-Phe-Arg	Bradykinin
Thr-Thr-Arg-Arg-Arg-Gly-Arg-Pro-Pro-Gly-Phe-Ser-Pro-Phe-Arg	Yellow jacket
Thr-Ala-Thr-Thr-Arg-Arg-Arg-Gly-Arg-Pro-Pro-Gly-Phe-Ser-Pro-Phe-Arg	Yellow jacket
Ala-Arg-Arg-Pro-Pro-Gly-Phe-*Thr*-Pro-Phe-Arg	Yellow jacket
Arg-Pro-Pro-Gly-Phe-*Thr*-Pro-Phe-Arg	Paper wasp
Pyr-Thr-Asn-Lys-Lys-Lys-Leu-Arg-Gly-Arg-Pro-Pro-Gly-Phe-Ser-Pro-Phe-Arg	Paper wasp
Gly-Arg-Pro-Pro-Gly-Phe-Ser-Pro-Phe-Arg	Paper wasp
Gly-Arg-Pro-Pro-Gly-Phe-Ser-Pro-Phe-Arg-*Ile-Asp*	Hornet
Gly-Arg-Pro-*Hyp*-Gly-Phe-Ser-Pro-Phe-Arg-*Val-Val*	Hornet
Ala-Arg-Pro-Pro-Gly-Phe-Ser-Pro-Phe-Arg-*Ile-Val*	Hornet

Figure 22. Bradykinin-like peptides.

Ile-Asn-Leu-Lys-Ala-Leu-Ala-Ala-Leu-Ala-Lys-Lys-Ile-Leu	*(Paravespula Lewisi)*
Ile-Asn-Trp-Lys-Gly-Ile-Ala-Ala-Met-Ala-Lys-Lys-Leu-Leu	*(Vespa xanthoptera)*
Ile-Asn-Leu-Lys-Ala-Ile-Ala-Ala-Leu-Ala-Lys-Lys-Leu-Leu	*(Vespa mandarinia)*
Val-Asp-Trp-Lys-Lys-Ile-Gly-Gln-His-Ile-Leu-Ser-Val-Leu	*(Polistes jadwigae)*

Figure 23. Amino acid sequences of mastoparans from different wasp species.

called "wasp-kinins." These are composed of 9 up to 18 amino acids and show pharmacological properties similar to bradykinin, which exists in the blood plasma of mammals. Bradykinin is a nonapeptide, released in the plasma by the action of special proteases, the so-called kallikreins. Bradykinin is a very potent agent, which decreases the blood pressure by dilating all the small blood vessels, the permeability of which is increased. It also has strong pain-producing properties when injected locally. Several different wasp-kinins from the genera *Vespa*, *Vespula*, *Paravespula*, and *Polistes* have been analyzed with respect to their amino acid sequences (Figure 22). Common to all is the 9 amino acid long sequence, which characterizes bradykinin. The wasp-kinins show physiological effects similar to bradykinin and are likely to be one of the several pain-producing factors in wasp venoms. Kinins have also been identified in the skin of various amphibians.

In wasp venoms there are also the so-called mastoparans, proteins possessing mast cell degranulating properties (Figure 23). These proteins have a unique chemical structure, but their mechanism of action is not fully known. It is different from that of the mast cell degranulating factor (MCD) in bee venoms.

Wasp and bee venoms contain some similar enzymes, for example, phospholipases and hyaluronidase, but otherwise have very different compositions. Among others, melittin and apamin are absent in wasps and mastoparans are absent in bees.

The often extreme pain-producing properties of wasp venom have deterrent effects on vertebrates. As for bee venom, the pain is mainly attributable to melittin. In the wasp venom the pain-producing factors are probably several, of which free histamine and serotonin can act directly on pain sensitive sensory nerve terminals. Other components of the venom may exert their effects indirectly, by causing the release of biogenic amines or other pain-producing substances from the tissue of the victim.

Ants (Formicoidea)

Ants are spread all over the world. They are social insects and are represented by about 15,000 different species. The number of individuals in some species is enormous. Therefore, considering the biomass, the ants may be the dominating predator amongst land-living animals. Certain species are very noxious animals due to their extreme plundering of plants, for example, the famous African driver ant (*Dorylus*) and the American legionary army ant (*Eciton*).

Generally, ants are omnivorous, eating a diet of both plants and animals. There are several aggressive ant species. In Australia, the primitive Jumper ants and Bull ants especially seem to be a problem. Their stings are painful, and can cause allergy which may be fatal in man. These and several other ant species use a barbless sting, situated on the tail, for injecting their venom. Other species, for example the formicine ants (*Formicinae*), have a reduced injection apparatus and instead spray the venom into wounds caused by their powerful front jaws or mandibles with which they grasp their victims. The common wood ant belongs to this kind of ant. The venom of formicine ants contains mainly formic acid.

Our knowledge of ant venoms is still very limited. One reason for this is the difficulty encountered in collecting venoms in sufficient quantities for chemical analysis. The compositions of the venoms of various ant genera often show large differences and do not permit any generalizations about their content. It was previously a common belief that all ant venoms contain formic acid. Today we know that formic acid is only present in one (*Formicinae*) of 11 subfamilies.

Myrmicine ants (*Myrmicinae*) represent the biggest subfamily among ants, with varieties all over the world. Among the species there is a wide range in the composition of the venoms. Besides various proteins, there are different alcaloids, monoterpenes, dialkylpyrrolidines, pyrrolines and piperidines. These complex hydrocarbons have a deterrent effect on predators and cause paralysis or death in smaller insects. The stings of some species are very painful for man. One example is the imported fire ant (*Solenopsis saevissima*), which has become a considerable problem in the southeastern United States. Its name refers to the burning pain it causes. The severe anaphylactic reactions to this venom appear to be due to its unique alkylated piperidine components.

Closer analyses of the complicated chemistry and pharmacology of the ant venoms have just started. From this research one can expect exciting new pharmacological tools and strategies for control of insect populations, which are based on natural biological principles.

Termites (Isoptera)

Termites are from several points of view a very interesting order of insects, represented by more than 2,000 species, mainly found in tropical parts of the world. Termites are commonly confused with ants, which is probably the reason that termites are sometimes referred to as "white ants." However, there are several distinct differences between these two groups of insects, and termites appear to be more closely related to the cockroaches. Termites are medium-sized insects. Mating termites are darkly pigmented, winged insects, while their offspring are wingless, white, often translucent and sterile.

Termites live in large colonies much like those of ants, and have several different social castes. The nests vary from simple cavities in the soil or wood to impressive complexes that project well above the ground. The construction of the nests is ingenious and involves ventilation systems, rain water drainage, maintenance of constant temperature and other adaptations that favor survival of the colony. The tasks of individual members of a colony are strictly

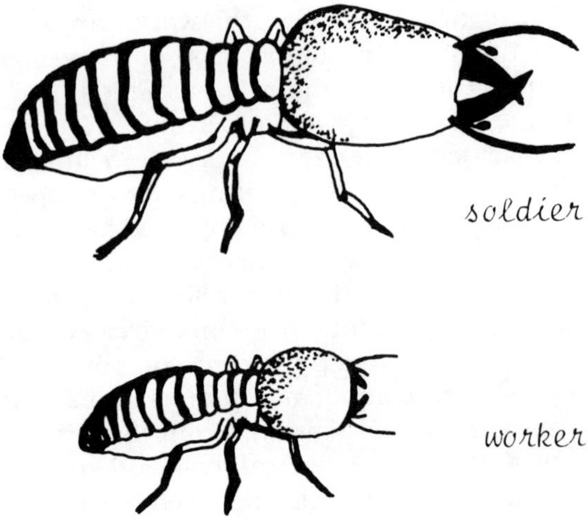

Figure 24. The life of the termite soldier is devoted to the defense of the nest.

coordinated to produce this complex structure, which sometimes is looked upon as a "superorganism" which grows, differentiates, attains a steady state, and reproduces.

Termites are usually divided into six families, of which the one richest in species, Termitidae, belongs to the "higher" and the other families to the "lower" termites. This classification is based on differences in social organization, morphology and other characteristics.

There are four principal castes: kings, queens, workers and soldiers (Figure 24). We shall concentrate on the soldiers. They devote their entire lives to the defense of the nest and the royal pair. Nature has with great inventiveness equipped the soldiers with a fantastic arsenal of chemical and mechanical weapons, this specialization taking place at the expense of other functions. Besides being sterile, the soldiers are blind and dependent on the workers who give them food. They compensate for their blindness with vibration sensitive receptors. With the help of these receptors the workers can localize attackers, often ants. Soldiers gener-

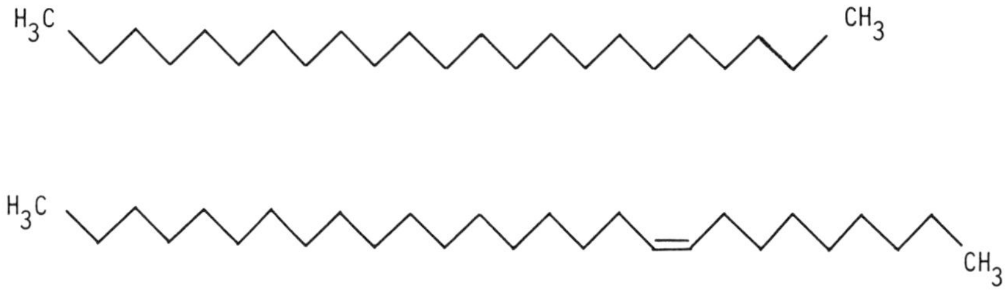

Figure 25. Simple hydrocarbons from termites.

ally have enlarged, sclerotized heads with well-developed mandibles. The larger part of the head consists of a venom producing gland and muscles by which the powerful mandibles can be maneuvered. The mandibles are specially designed to bite, chew and to sting the intruders. In certain species the mandibles are even equipped with saw teeth. The construction of the mandibles, and more recently the composition of the venoms, contribute to the taxonomic classification of termites.

Different termite species have developed different strategies for the chemical defense of the termite colony. Three main methods can be distinguished. The first one is biting in combination with the application of a relatively simple venom in the wound caused by the strong mandibles of the soldier. The second is daubing, using an enlarged upper lip (labrum) which resembles a paintbrush, to apply a contact venom to the cuticle of the intruder. In the third and most advanced method of defense, a complicated venom is squirted at the undesirable visitor.

The first method, where a wound is poisoned, is used by several different termite species and is also common among the "higher" termites. This type of defense has evolved independently several times and may be regarded as an example of convergent evolution. In certain species, the venom is very simple, consisting of long-chain saturated or unsaturated hydrocarbons (21-35 carbon atoms long), so-called alkanes and alkenes (Figure 25). These substances are not toxic as such, and lack effect if they are applied to an intact insect. However, if the cuticle of the aggressor has been punctured by the powerful mandibles of the termite, the hydrocarbons will exert harmful effects, by softening the punctured area

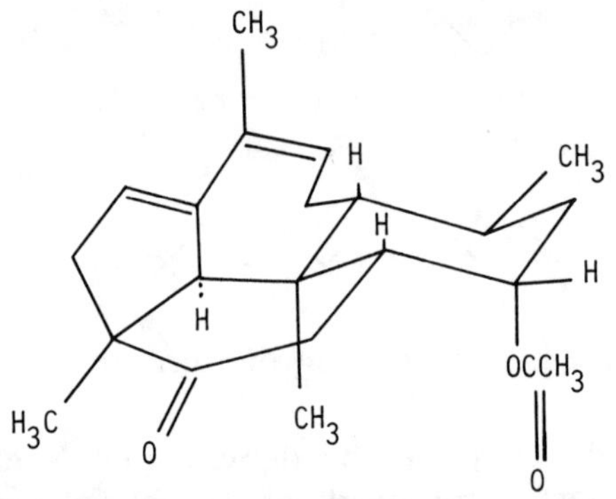

Kempane

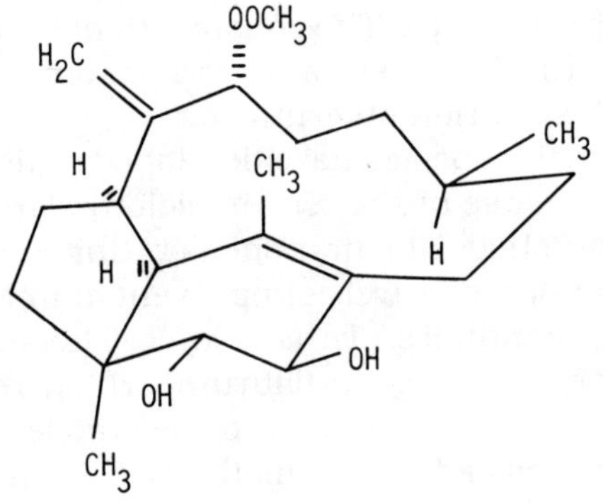

Trinervitane

Figure 26. Cyclic diterpenes are unique substances from termites.

and preventing its healing. The molecules also prevent coagulation of the insect's hemolymph.

More complicated and toxic hydrocarbons have also been found in the venom of some biting species. These hydrocarbons, cyclic diterpenes, are unique substances, which no other animals or plants are known to synthesize (Figure 26).

In the second main method of defense, extremely oil-soluble sub-

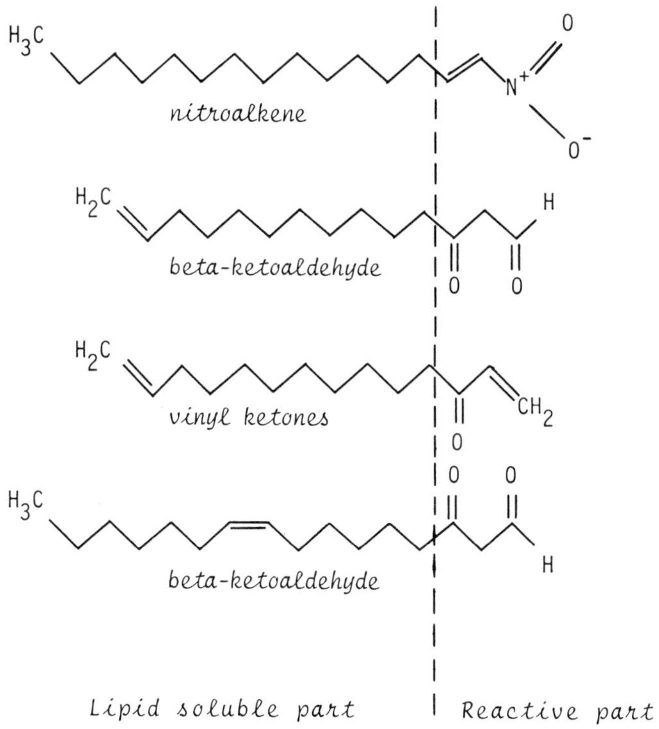

Figure 27. Lipid soluble and very reactive termite toxins.

stances are spread over the intruder with the paintbrush-like lip. The substances (e.g., nitroalkenes, β-ketoaldehydes and vinyl ketones) are derived from fatty acid molecules and have the ability to penetrate the cuticle. In one end of the chain, the molecule has a strongly reactive group that has oxidizing effects which cause the toxicity (Figure 27). Such chemical groups often occur in defensive secretions, for example, in those of some fungi and higher plants. They are known to inhibit cell growth and to be insect repellents. These toxic hydrocarbons are produced by a large gland in the head of the termite and are also stored in a reservoir in the abdomen. The total weight of the toxins can amount to 30% or more of the dry weight of the soldier and one may ask how the termites avoid being poisoned themselves. The existence of a special enzyme system appears to be the answer to this question. If the toxins should appear in termite tissues, enzymes reduce the reactive groups and transform the molecules to less toxic products.

The third and most advanced of the three main methods of chemical defense has evolved in the family, Nasutitermitinae. The name refers to the long and narrow snout, the nasus, which is characteristic of the members of this family. The toxic and sticky secretion is produced by a frontal gland which connects to the snout. The gluelike venom is squirted from the snout against the intruder. The very viscous solution consists of complex toxic diterpenes (e.g., secotrinervitane, kempane, rippertane and longipane) dissolved in simpler monoterpenes. The diterpenes have extremely amphiphilic—that is wetting—properties due to the presence of both hydrophilic and hydrophobic groups, which facilitate their penetration through the very hydrophobic cuticle of the intruder. Due to its gluelike quality the viscous secretion also exerts mechanical effects and renders the movements of the attacker more difficult. In its efforts to get rid of the secretion, the secretion will instead be spread over a large part of the body, increasing its deleterious effects.

The most troublesome enemies of the termites are probably ants. Among the predators one also find several mammals, such as the anteaters, ant marsupials and ant bears. The mechanisms behind the resistance of these animals to the venomous termite soldiers are still unknown.

Butterflies (Lepidoptera)

Butterflies form the second largest order of insects, with more than 165,000 species known. Several species remain to be discovered and classified.

Among the butterflies, it appears to be the moths (Bombycids, Noctuids, Sphinx-moths) that produce toxic substances in chemical defence against predatory animals. The larvae are more frequently venomous than the adult butterflies. The larvae of venomous species are mostly very hairy, and special hairs or setae are fortified with venom originating in an attached miniature gland. The setae are broken at the slightest touch of the attacker and stick to his skin. The larvae are easy prey for birds and other smaller predators and against them the toxic substances can act in passive chemical defense. Irritating skin reactions may also occur in humans by contact with the hairy and venomous larvae of some moths. In connection to cyclic population peaks of certain species there may arise medical problems due to skin inflammations and allergic symptoms.

Detailed investigations of the toxic substances of these animals are still very limited. The venoms seem to contain mainly proteins, and among them various enzymes, such as, proteases, esterases and phospholipases. In contrast to several other insect venoms there are relatively small amounts of biogenic amines present.

Bugs (Heteroptera - Hemiptera)

The external appearance of bugs is often similar to beetles. They are found in most places of the world, most frequently in warm and temperate regions. The number of species is usually stated to be about 25,000. There exist terrestrial (e.g., the chinch bug and the harlequin bug) as well as aquatic bugs (e.g., the skipper and the backswimmer).

Several terrestrial and most aquatic species are equipped with very effective venomous offensive weapons which also are used to prey on other insects, and even on small amphibians and fishes. These bugs often have a well developed upper jaw, sometimes armed with sharp-pointed teeth able to puncture the prey. Venomous secretions can then be applied to the wound with the use of hollow stilettos attached to the lower jaw.

The toxic substances are produced by the salivary glands. A venom mostly consists of numerous toxic proteins and in one species 13 different proteins have been identified. The toxins appear to affect a wide range of physiological functions, including neuromuscular transmission. The still poorly characterized venoms contain, for example, hyaluronidase, proteases, esterases, lipases, cytolytic and neurotoxic substances. By a simultaneous action on several different mechanisms the conditions increase to cause a rapid and effective immobilization of the victim (compare Cone snails). This is essential because then the prey does not have a chance to escape the perhaps somewhat slower predator.

One interesting example is the African bug (*Platymeris rhadamanthus*). Its venom exerts an almost instantaneous toxic effect. Instead of injecting the venom this species is able to spit its venomous saliva for a distance up to 30 cm, and with relatively good precision. If a cockroach happens to be the target, it takes only 3 to 5 seconds until the animal is paralyzed. The potency of bug venoms is also illustrated by the finding that the venom of a single specimen of another species (*Holotrichius innesi*) is enough to kill a mouse within 30 seconds.

Bites of certain species can be very painful for humans. Assassin bugs are sluggish predaceous insects which somtimes attack man, inflicting a burning wound. The pain is very severe, may affect a large part of the body and remain for weeks.

Most terrestrial bugs are equipped with stink glands, which during the larval stage are located on the back, and in the adult form along the sides or under the animal. These glands produce volatile, often extremely nasty-smelling, but not very poisonous substances. The secretion serves defensive and/or signal functions.

Beetles (Coleoptera)

This is by far the largest order of insects, and therefore the largest order in the animal kindom. Beetles comprise more than 300,000 described species. The majority of beetles are terrestrial and they are to be found in most places of the world.

In contrast to bugs, beetles appear to lack offensive toxins. However, the existence of defensive venoms or secretions is probably

Figure 28. Bombardier beetle.

common in this successful group of animals, the toxinology of which is poorly known.

Among beetles, a particularly odd and impressive chemical defensive technique is found in the bombardier beetles (*Brachinus sclopeta*), which has attracted much interest during the past few years (Figure 28). These beetles eject a hot spray from glands located in the posterior part of the abdomen. There are two kinds of glands, an inner gland connected to an outer gland, which in turn is connected to a narrow opening in the abdomen. In the inner gland a mixture of hydroquinones and hydrogen peroxide is produced and in the outer reservoir special enzymes, peroxidases and catalases, are synthesized. When a bombardier beetle is attacked, the content of the inner gland is squeezed into the outer gland, mixed with the enzymes, and ejected under high pressure as a fine spray, an explosive gas, which can be aimed at the attacker. When the different substances are allowed to come into contact with one another, oxygen is liberated as a result of the mixing of catalase and hydrogen peroxide and the hydroquinones are oxidized to quinones under the very strong generation of heat. The detonation is even audible. The temperature may reach 100°C and this procedure can be repeated in quick succession. The beetle can aim the abdominal tip in virtually any direction from which it is attacked. The intruder rapidly looks for a safer place with a more pleasant temperature and in the future avoids these highly dangerous dynamitards.

Also legendary is another beetle, namely the Spanish fly (*Lytta vesicatoria*), found throughout southern Europe (Figure 29). A substance called cantharidin is derived from its body. This substance has been described as having the ability to arouse an increased sexual desire in man, to be an aphrodisiac. This idea is probably based on the fact that cantharidin acts as a strong irritant of the urogenital system and can trigger a convulsive and painful erection of the penis. A feeling of sickness, stomach cramps and renal failure are other harmful effects of this substance. There are even reports that the Spanish fly has caused some deaths in humans. The role of cantharidin in beetles is unknown. It may be intended as a defense against insect predators.

The Spanish fly is a very well-known insect. Less familiar but considerably more dangerous are certain beetles that eat leaves. They are found in Africa and their venoms are used as arrow poisons. Bushmen, hunter and gather peoples who inhabit the Kalahari desert of Botswana and eastern Namibia, use the body contents of pupae of certain species to poison the arrows needed for hunting big animals such as the giraffe, zebra, wildebeest and others. Of the different beetle species used, the most important one seems to be *Diamphidia nigro-ornata*. The Bushmen themselves call this beetle "Nga" or "Ngwa." The Bushmen dig up the pupal cocoons and squeeze the body fluids directly onto the shaft behind the tip of the arrow head. The sharp arrow tip itself is not coated, for fear of accidental poisoning while handling the arrows. The extracts of about 10 pupae make ammunition for one arrow. The besmeared arrows are carefully dried over hot coals and the poison retains its toxicity for up to a year. The arrows are discharged with small bows at close range from the victim. For large animals like giraffes, death may come first after some days. However, the ability of the poisoned animal to move is rapidly reduced and the life of the prey can therefore be put to an end at an early stage with other weapons, for example spears.

The toxic component of the extract of this beetle is a protein

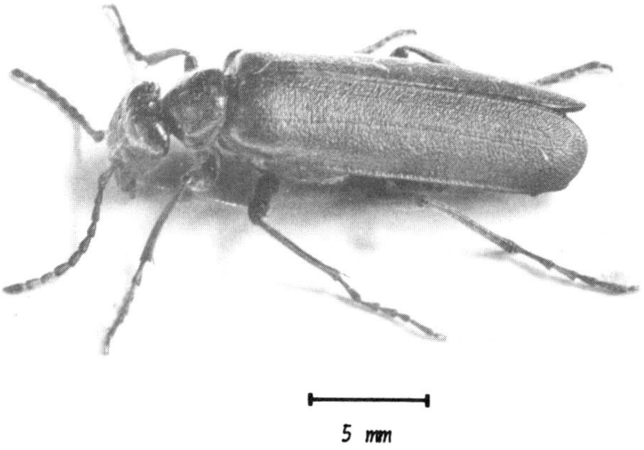

Figure 29. Spanish fly.

with a molecular weight of about 54,000. It is extremely potent and a few μg of it is enough to kill a mouse at i.p. injection. The mechanism of action is not still elucidated. Some observations suggest that it has a depolarizing effect on muscle cells and probably also on other cell types. It could be that the toxin causes a general increase in cell permeability for small ions. Some investigators believe that the main effects of the toxin can be attributed to hemolysis and thus reduced oxygen-carrying capacity of the blood.

The role of the toxin in the pupae is intriguing. The pupa is well protected during its life by the cocoon and a deep layer of sand. The biology of the arrow poisons used by the Bushmen seems to be an exciting and still unexplored field in toxinology.

13

Insect Use of Plant Poisons

There are many examples of plants which protect themselves against herbivores, especially insects, by means of various toxins which poison the attacker or impair its ability to digest the plant. The poison may also have unfavorable effects on the reproductive capacity of the predator. The strategies for defenses are numerous and can in part explain the great number of plant toxins. However, these chemical defenses have been exploited by several insect species for their own advantages. In coevolution with poisonous plants, insect species have developed, which possess specific resistance towards certain poisonous plant species and even the ability to store the toxins and use them for own defense against predators.

The common aspen tree contains salicin, a toxic derivative of phenol, which among other things inhibits oxidative metabolism. In this way the aspen is protected against most, but not all, herbivores. Among the latter there is a special beetle, which has developed tolerance against the poison and has even learned to transform salicin from the aspen into a nasty smelling and possibly toxic component, stored in glands along the back of the beetle. In response to danger, the substance is secreted and discourages most predators from closer contact with the beetle.

Another interesting example is the grasshopper, *Poekilocerus*

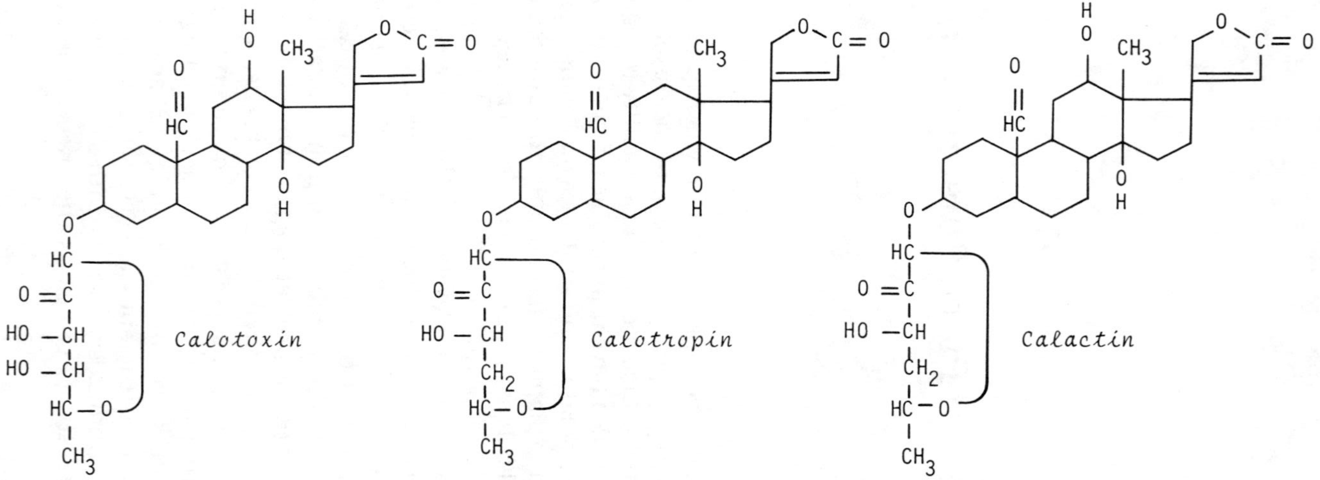

Figure 30. Cardenolides are toxins from milkweed species, which certain insects use for own defense.

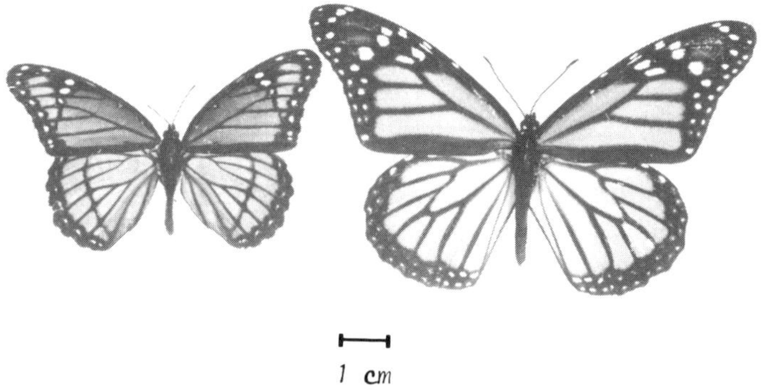

Figure 31. Viceroy butterfly (left) gets protection by a similar coloration and pattern as the monarch butterfly.

bufonius, which feeds solely on plants of the poisonous milkweed family (Asclepiadaceae). These plants manufacture several complex glycosides, known as cardenolides (calactin, calotropin, calotoxin, etc.) (Figure 30). Cardenolides refer to their very toxic effects on cardiovascular functions. These substances are related to digitoxin, which is found in the poisonous foxglove, *Digitalis purpurea*. Like digitoxin, the cardenolides inhibit the activity of the enzyme Na^+,K^+-ATPase, which leads to several different physiological effects, for example, a reduced heart rate. The grasshopper can store the cardenolides in a special gland and when it is attacked can defend itself by ejecting a spray of the poison against the enemy.

The larvae of Monarch butterflies also feed on the milkweed species, and especially on the common milkweed, *Asclepias curassavica*, which grows wild in large areas of South and Central America. Milkweed poison is stored in the body fluids of the larva without any noxious effects on either the larva or the adult butterfly (Figure 31). In contrast, the bird which believes it has found a tasty butterfly will soon regret the mistake. Its rescue is that the cardenolides, at a concentration lower than the lethal one, stimulate the nerve center in the brain that controls vomiting, which results in a very unpleasant gastric upset. The bird will never try again. Other birds, which have passively witnessed the drama, probably learn from the faux pas of their companion and in the future avoid all Monarch butterflies. The potency of the poison is illustrated by the fact that the extract of a single butterfly is enough to kill a cat or a small dog. Some nonpoisonous butterflies and their larvae have taken advantage of the situation by assuming a coloration and pattern which closely mimics the poisonous species. They make their way by bluffing a presumptive predator. Defense which makes use of a protecting disguise, mimicry, is common in the animal kingdom and not least among insects.

14

Spiders and Scorpions (Arachnida)

The goddess of handicrafts, Minerva, transformed the maiden Arachne into a spider because Minerva feared Arachne as a competitor in the art of weaving. In this way Arachne's name has been linked to the class of animals (Arachnida), which includes not only spiders (Araneida) but also scorpions (Scorpionidea). The arachnids are arthropods, but have probably at an early stage of evolution formed their own group, which now is distinct from the rest of the arthropods in several respects.

Spiders (Araneida)

There are at least 30,000 spider species throughout the world, and new ones are regularly being found. The true number of species may in fact be 100,000 or more. They are mainly terrestrial and only some species have adapted to aquatic life.

Spiders have always fascinated man and often play an important role in folklore and in various ancient legends. This is not surprising, considering the often bizarre exterior of spiders, with

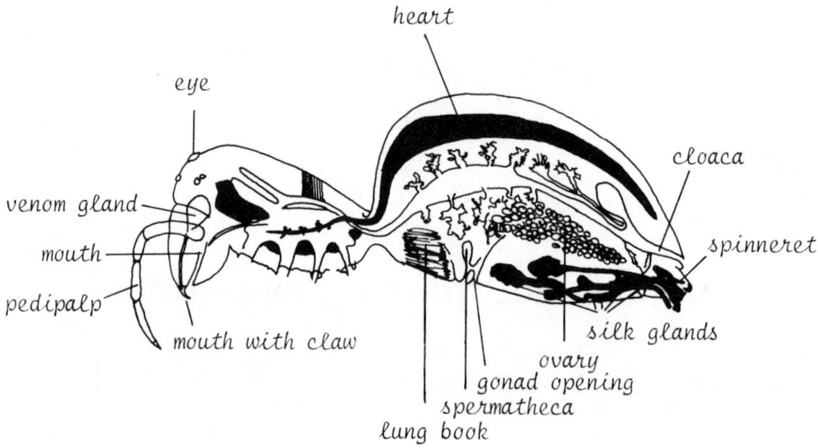

Figure 32. The anatomy of the spider.

their various processes, often covered with hair, eyes situated on shafts, large abdomen, ability to spin webs and not the least, their venomous properties. In particular the latter has certainly been a reason for the reaction of fear and disgust so many people have to spiders, even though they are mostly harmless and useful organisms which prey on various insects, for instance harmful ones.

Spiders generally have a venom gland in the cephalothorax, which is connected to a claw in the first pair of appendages, the chelicerae (Figure 32). The tip of the claw contains the opening from the venom gland. When the claw is pressed into the prey, the poison is allowed to run freely into the open wound. The second pair of appendages, the pedipalps, has a sensory function and is also a copulatory organ in the mature male. The last four pairs of appendages are walking legs.

Spiders are extremely carnivorous and feed mainly on insects for which their venoms are designed. A few species are considered to be of significant danger to man.

Spider venoms represent kinds of naturally existing toxic compounds, which are of great pharmacological interest. These venoms, like several other toxins of animal origin, exert their action primarily on the peripheral nervous system. The toxins consist of high molecular weight components which can not pass the blood-brain barrier. Like the venoms of several other arthropods, spider venoms also contain various biogenic amines. The two best known spider genera from the toxinological point of view are *Latrodectus* (Widow Spiders) and *Loxosceles*, which both can cause serious envenomation in humans.

Of the *Latrodectus* species, the Black Widows in particular have attracted much scientific interest. There is one species in southern Europe (*Latrodectus mactans tredecimguttatus*) and one in the United States (*Latrodectus mactans mactans*). The venoms of their close relatives, the Brown Widow and the Red-back spider, give rise to similar neurotoxic effects to those of the Black Widow.

Black Widows are often found in cornfields; harvesters may be attacked. These spiders are small and can be difficult to detect. They are black colored with red spots—nature's warning—on the abdomen. Mature females are about 14 mm in body length. They are much bigger and contain more venom than the males. The volume of the venom appears to undergo seasonal variations, at least in the American species, being more dangerous in the autumn than in the spring. The syndrome caused by Black Widows is known as latrodectism. The venom affects the motor and sensory nerve terminals of the victim. The bite can pass unnoticed for several minutes but may be felt as a shooting and localized pain. After that, the pain intensifies and spreads to other parts of the body. Severe paroxysmal muscle cramps may occur. The patient becomes anxious, shivers and complains about difficulty in breathing and sometimes has a fainting feeling. These symptoms reach maximal intensity 6-12 hours after the bite, and the whole process is very painful. Approximately two days after being bitten the symptoms start to subside. In spite of the dramatic reactions, latrodectism is seldom lethal. However, for elderly people or children, the consequences may be serious. Fortunately antivenoms—antibodies—towards various spider venoms, for instance the Black Widow's, are available today.

Venom gland extracts intraperitoneally injected in rats are 15 times more potent than the venom of the Prairie rattlesnake

(*Crotalus viridis*). However, since the amount of venom released by a spider bite is very small, the toxic effect is generally limited. In the Mediterranean area, venomous spider bites are not always associated with Black Widows, but often with the mildly venomous and rather harmless tarantula.

Like most other spider species, the Black Widow captures its prey, usually an insect, with the use of a silken web. While the insect tries to get rid of the web it sets up vibrations, which the spider registers by tactile hairs on its legs. The spider, often hidden in some corner of the web, quickly approaches its prey and secures it by entangling it in newly woven sticky threads. Thereafter the venom is injected into the victim which is immediately paralyzed. The Black Widow then applies drops of saliva to the prey. This saliva contains proteolytic enzymes, which slowly disintegrate the prey, which can later be sucked up by the spider. Nothing but the cuticula will remain of the insect.

The toxic component of the venom is a protein having a molecular weight of about 125,000, and is called latrotoxin or BWSV (Black Widow Spider Venom). The exact mechanisms of action are not fully known. It exerts pronounced morphological and physiological effects on most types of peripheral nerve terminals, irrespective of their transmitter type, causing a massive transmitter release and degeneration of the terminals. The effect is limited to the very nerve terminals and no effect has been observed on the more proximal parts or on the surrounding Schwann cells. The effect of latrotoxin has mainly been studied with the use of nerve-muscle preparations from frogs and mammals. In these systems, the toxin causes a rapid release of vesicular acetylcholine, followed by transmission inhibition. The nerve terminal becomes swollen and shows a disintegrated cytoskeleton, damaged mitochondria and empty vesicles. It is believed that latrotoxin forms an ion channel for Ca^{2+}, Na^{+} and K^{+}. An increased inflow of Ca^{2+} and other cations triggers an enhanced release of the transmitters. The possibility has also been discussed that latrotoxin facilitates fusion of the vesicles and the presynaptic membrane. It is possible that the latter is a consequence of the increased intracellular Ca^{2+} concentration, which stimulates the activity of certain lipases and neutral proteases of importance for the fusion process. Latrotoxin is an interesting experimental tool to study nerve-muscle transmission and the healing processes taking place after latrotoxin has induced denervation close to the muscle.

Several scientists believe that species of the *Loxosceles* genus are a larger public health problem than Black Widows. *Loxosceles* species are of similar size or slightly smaller than Black Widows. There are over 100 species of *Loxosceles*, widely distributed in the world. Twenty of these species are found in Africa and one in the Mediterranean area. Most species are known from North, Central, and South America, and the West Indies. The bites of certain species are very painful. The most dangerous species exists in Uruguay and in Chile and it has caused the death of several humans.

The brown recluse spider, *Loxosceles reclusa*, is a medically important spider found in the southwestern part of the United States. Some human deaths have been attributed to bites of this spider. The most common response is a local necrotic lesion, although the bite itself is usually not particularly painful and individuals therefore often ignore it. However, the bite can start a prolonged inflammation of the skin, and if this becomes infected, there may be serious consequences resulting in fever, muscle weakness, nausea, and vomiting.

There is a comparatively large literature on the chemistry and toxicology of the venom of *Loxosceles reclusa*. This venom contains both high and low molecular weight toxic proteins. The toxins are very lethal for insects, the natural prey of the spider, but have less and varying effects when injected in different rodents. One protein with a molecular weight of about 31,000 appears to be responsible for the necrotic lesions in humans.

The funnel-web spider (*Atrax robustus*), found in southeastern Australia, is one of the most dangerous spiders in the world and may also be lethal to man. The female, about 4 cm in length, is much bigger than the male. However, the male spider is much more toxic than the female and the male is also responsible for all reported serious envenomations in Australia when this spider is involved. Both sexes are very aggressive, which has contributed to their disrepute. Strangely enough, the funnel-web spider venom has relatively little effect on non-primate mammals. For example, rabbits can be injected with doses almost lethal to man without showing any serious symptoms. The reason for this appears to be the presence of an inhibitor of the toxin in vertebrates other than primates. In man, envenomation by this spider is painful and leads to circulatory and respiratory failure, muscle fasciculation, sweating, nausea, and vomiting. In severe cases, it can lead to shock and coma due to brain damage. Fortunately, an antivenom has been available for several years, which gives the victim a good chance of recovery if he can reach a hospital.

One toxic component of the funnel-web spider venom is a protein called atraxin, having a molecular weight between 15,000 and 20,000. It has not yet been purified and characterized chemically. Like latrotoxin, atraxin causes a massive release of neurotransmitters in the peripheral nervous system. In contrast to latrotoxin, atraxin does not induce any structural changes and its effects are reversible. Some observations suggest that atraxin activates the Na^+-channels, so that the threshold for generation of action potentials is reduced. However, the magnitude of the membrane potential is not affected. An increased tendency to spontaneous electrical activity induced by atraxin may therefore explain the enhanced release of neurotransmitters. This toxin is again an example of one of several unique biological tools which nature has created and which scientists can use in their laboratories to analyze basic and important physiological mechanisms.

The most poisonous of all spiders are Brazilian spiders belonging to the genus *Phoneutria*. These spiders are big (4-9 cm long) and aggressive. They can bite repeatedly if provoked and are able to kill small rodents and birds. They are also dangerous to man and their bites cause intense local pain, hallucinations, spasms and, in severe cases, respiratory failure. Luckily, antivenom is available and can prevent a fatal outcome. The mechanisms of action of the venom are not fully understood. The most potent toxin shows neurotoxic effects and similarities to certain scorpion toxins. It probably causes a delay in the inactivation of the Na^+ channel and also induces morphological abnormalities of peripheral nerves.

Scorpions (Scorpionidea)

Scorpions represent one of the oldest (425 million years) living groups of animals; they have retained many primitive characteristics although they are highly specialized for their mode of life. Scorpions are the largest of the arachnids and are, because of their poisons, very well known. They reach a length which may vary from 1 cm up to about 20 cm. The forepart of the body, the thorax, is broad and connected to a thinner abdomen which has an elongated tail equipped with a venomous sting. There are about 1,400 known species distributed over six families. Most species are found in the warm regions of the world, and they are only active during the night. Scorpions are shy animals with a plain display of colors, and are therefore difficult to detect despite their numerous occurrence in certain parts of the world, for instance, in the tropics. Some species have adapted to temperate areas like the Mediterranean countries.

The venom is produced by two sac-like glands, which are both connected to the sting, the feared weapon of the scorpion. Each gland is situated between a muscle and the hard outer skeleton. By means of muscle contractions the venom can be forced out into the sting when a prey is attacked. Like crayfish, the scorpions also have a pair of very strongly developed prehensile organs, chelate pedipalps, connected to the mouth region. Using these organs, the prey can be grasped when a lethal venom injection is delivered. All scorpions are venomous. A small number of species are potentially dangerous to man (*Buthus*, *Androctonus*, *Leiurus*, *Parabuthus*, *Mesobuthus*, *Hemiscorpion*, *Centruroides*, *Tityus*). In the United States, the only two dangerous species are found in southern Arizona (*Centruroides sculpturatus* and *Vejovis spinigerus*), whereas the many other species found in this country are relatively harmless to man. The largest number of envenomations take place in Mexico, where about 134 scorpion species or subspecies have been found, but only eight species are dangerous, and they are responsible for the majority of the stings. Some 100,000 stings occur each year and about 1,000 people die as a result. The fatalities tend to be children and the elderly and are primarily due

Table 2. Lethality in mice of venoms from various species of scorpions. After Watt and Simard, 1984.

Species	Range	LD_{50} mg/kg*
Androctonus australis	North Africa	0.32
Androctonus amoreuxi	North Africa	0.75
Androctonus mauritanicus	North Africa	0.31
Androctonus crassicauda	Turkey, Iraq	0.40
Buthus occitanus tunetanus	North Africa	0.90
Centruroides limpidus tecomanas	Mexico	0.69
Centruroides santa maria	Central America	0.39
Centruroides sculpturatus	Southwest U.S.A.	1.12
Leiurus quinquestriatus	Mideast, Israel	0.25
Parabuthus transvalicus	South Africa	4.25
Tityus serrulatus	Brazil	0.43

*Venom given subcutaneously

to species belonging to the genus *Centruroides.* LD_{50} values for the venoms of these and some other dangerous species are about 0.5-1 mg/kg body weight using subcutaneous injection in mice, i.e., values associated with high toxicity (Table 2).

The symptoms of scorpion stings range from local pain to very complex neurotoxic and cardiovascular effects and, if it comes to the worst, respiratory arrest and death. The pain which most scorpion stings produce, is due to serotonin and possibly an additional factor. The subsequent symptoms can be ascribed to different toxic proteins, which are of great scientific interest and which have been subjected to extensive investigations. There is no entirely satisfactory treatment of envenomated people. Scorpion toxins are not good antigens and do not produce a very potent antivenin. In addition, toxins of one species do not necessarily induce antibodies which cross protect against toxins of other species. Therefore there is no universal antivenin available. Anyhow, antivenin in combination with other treatment is usually the recommended management of scorpion stings.

Venom for chemical and pharmacological studies is usually collected by electrical stimulation of the venomous sting. The venom is complicated and consists of a large number of components. Besides salts and biogenic amines, for instance, serotonin, up to 10 to 15 toxic proteins have been identified in the venom of *Centruroides* species and at least 25 toxic factors in that of the African species, *Buthus eupeus.* The proteins are generally basic with molecular weight of about 8,000 or less, and appear to function primarily as neurotoxins. About 80 different scorpion toxins are known with respect to the amino acid composition and the complete amino acid sequences of about 30 toxins from various scorpion species have been determined. The different toxins show similarities and form a family of homologous proteins. Toxins from the New World species resemble one another more than they resemble toxins from the Old World species. The tertiary molecular structure (3-dimensional structure) has been elucidated for one toxin (Toxin Var3) of the American species, *Centruroides sculpturatus* (Figure 33). This structure is regarded as typical of scorpion toxins in general. Four disulfide bridges stabilize the structure and make the toxins extremely resistant to denaturing factors, changes in pH, high temperature, etc. However, the toxins are inactivated by reagents which break the disulfide bridges.

Some scorpion toxins are more lethal to invertebrates than to vertebrates, and vice versa. There is also a certain species specificity among the different toxins, which is reflected in the designation of the toxins, for instance, mammal toxin, insect toxin, crustacean toxin, etc. It has long been known that scorpions themselves are very resistant to scorpion stings.

Several scorpion toxins exert their action via the voltage dependent Na^+ channels in excitable cells. Based on these effects one may distinguish three general classes of toxins. The first class delays the inactivation of the Na^+ channel, which results in a prolongation of the action potential to many times its normal duration. Toxins showing these properties appear to exist in venoms of most scorpion species. Toxin Var3, the tertiary structure of which is known, belongs to this class of toxins. The second class of toxins affects the activation of the Na^+ channel. These toxins have only been identified in scorpions from the New World and cause a transient shift in the voltage dependence of activation of

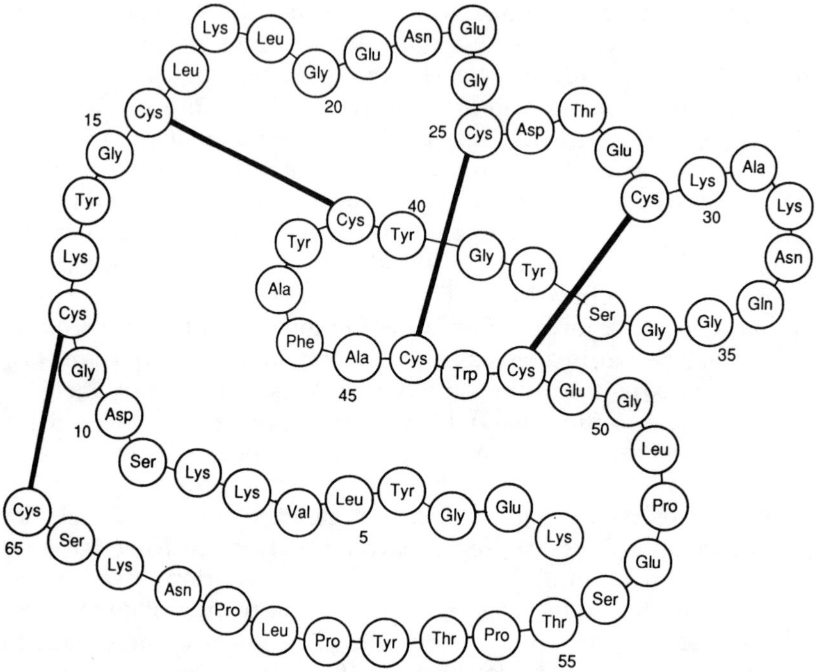

Figure 33. Toxin Var3 of an American scorpion, Centruroides sculpturatus.

the Na^+ channel. The result is an increased tendency of the membrane to fire spontaneously and repetitively. The third class of toxins reduces Na^+ as well as K^+ currents, with no changes in the activation or inactivation mechanisms.

However, scorpion toxins are far more complicated than this classification suggests. For example, several recently characterized scorpion toxins have been shown to affect different types of K^+ channels specifically. Examples of such toxins are charybdotoxin (*Leiurus quinquestriatus*) and noxiustoxin (*Centruroides noxius*). The availability of such toxins is proving invaluable in characterizing the many subtypes of the K^+ channel.

Research on the mechanisms of action of scorpion toxins is an excellent example of a field where biochemists and electrophysiologists successfully work side by side. As a result, new pharmacological tools have been found which have contributed to understanding the mechanisms that underlie the diversity of cellular Na^+ and K^+ channels.

SELECTED COLOR ILLUSTRATIONS

FROM THE FILES OF

FREDRIC L. FRYE, D.V.M, M.S.
Fellow, Royal Society of Medicine

Sea anemones

Female black widow spider, *Latrodectus mactans*

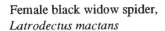

Female black widow spider, *Latrodectus mactans*, with egg mass

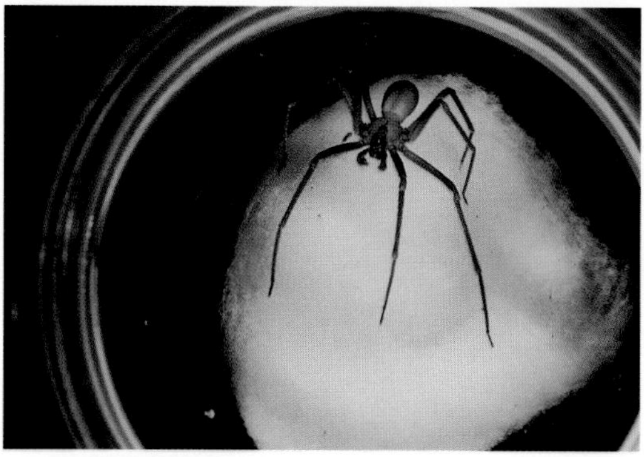

Male brown recluse spider,
Loxosceles reclusa

Female emperor scorpion,
Pandinus imperator

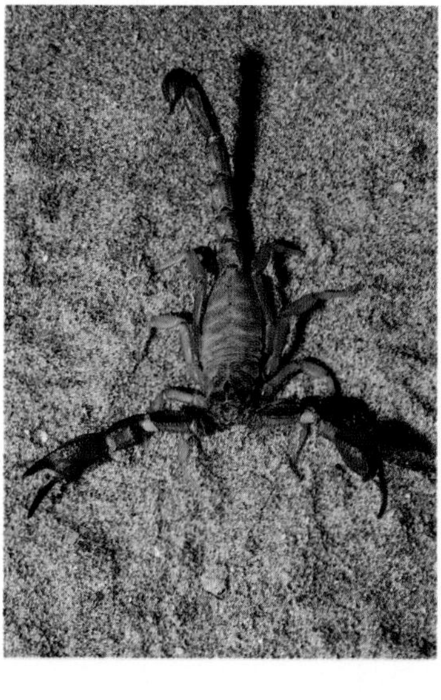

Scorpion, in defensive pose,
sex and taxonomy unknown

Two horned sand vipers, *Cerastes cerastes*, partially buried in sand

Baja California rattlesnake, *Crotalus enyo*

Timber rattlesnake, *Crotalus horridus*

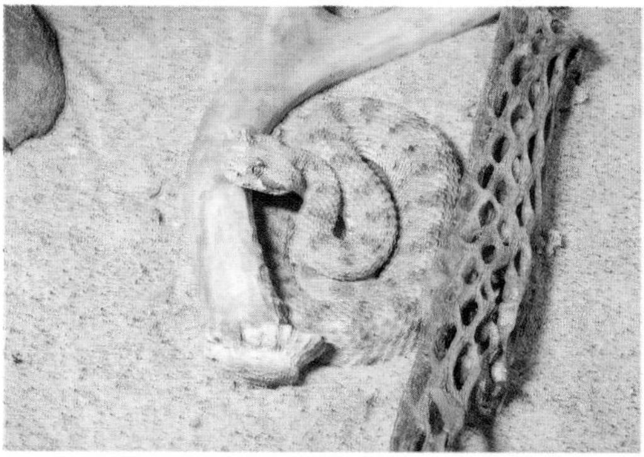

Sidewinder rattlesnake,
Crotalus cerastes

Head of cottonmouth water mocassin, *Agkistrodon piscivorus*

Two copperheads, *Agkistrodon contortrix*, and a cottonmouth water moccasin, *Agkistrodon piscivorus*

Cobra enclosure at the
Red Cross Venom Institute,
Bangkok, Thailand

Mexican beaded lizard,
Heloderma horridum
and Gila monster,
Heloderma suspectum

Western toad, *Bufo boreas*

Poison arrow frog,
Dedrobates sp.

European honeybee in flight
with full pollen baskets

Poisonous dragon fish,
Pterois sp.

15

Fishes (Pisces)

Fishes are poikilothermic animals belonging to the vertebrates, a subphylum of the chordates. They have gills for respiration and fins for swimming. When it comes to number of species, fishes are outstanding among the vertebrate groups. There are about 20,000 reported fish species, but new ones are still being discovered. About 60% of the now existing species are marine, and the rest freshwater animals. Fishes (Pisces) represent a "superclass" usually divided into the classes cyclostomes (50 species, e.g., lampreys and hagfish), elasmobranchs or cartilaginous fishes (550 species, e.g., sharks and rays), and bony fishes (19,500 species, e.g., perches and cod).

At least 1,000 fish species are known to be poisonous, and of them, a large number are venomous, having specialized cells for producing toxins and a structure for delivering the venom. There are several poisonous fishes, distinguished from venomous fishes, which have no such apparatus but still have toxic tissues. Venomous fishes are sometimes called "phanerotoxic," and those with toxic tissues "cryptotoxic" fishes. In the latter case, the toxicity can be ascribed to the ingestion of poisonous dinoflagellates or other toxic organisms. Some of these fishes have tissues that are toxic at all times, whereas others are poisonous only at certain periods or in special regions. In addition, the toxicity can sometimes be restricted to certain tissues. Toxins of "phanerotoxic" and "cryptotoxic" fishes show no chemical or pharmacological relationships.

"Phanerotoxic" fish toxins vary notably in their chemical and toxicological properties. Some venoms cause rather simple effects, such as transient vasoconstriction or dilation. Other venoms produce more complex responses, such as disturbances in the parasympathetic function and in the blood circulation. However, there appears to be some toxicological similarities between different "phanerotoxic" fish venoms, and there is one remarkable and common property of these venoms. They all cause an immediate, intense pain. The pain is usually far more severe than that produced by a snake or a Black Widow bite. The quality of the pain suggests that one or more factors are similar in venoms of, for instance, stingrays, weevers, stonefishes and stargazers.

16

Phanerotoxic Fishes

About 200 fish species, equipped with a well-developed venom apparatus, have been described. Among the more well known are stingrays, spiny dogfish, weevers, zebrafishes, stonefishes, waspfish, scorpionfish, stargazers and some catfishes, sharks, ratfishes, carangids, and surgeonfishes. Most venomous fishes are slow swimming, non-migratory animals which are often found in shallow waters, around rocks, corals or buried in sand.

The venom apparatus is used mostly for defensive purposes. The venom glands are connected to special stings or spines which inject the venom when the spines touch or penetrate a surface. The spines are usually associated with the gill covers and dorsal, pectoral, or anal fins of the fish. The spines are also often mechanically injurious to the victim and the wounds facilitate the spreading of the venom. In certain species the surface of the spines consists of venom producing cells, which can remain in the victim for some time exerting toxic effects.

We know comparatively little about the venoms of the various species, probably due to the often small amounts of venom which can be extracted from the venom glands, and also due to the instability of the lethal components. The pain-producing factors are probably low molecular and rather stable substances, whereas the dangerous toxins are proteins with molecular weights between 50,000 and 800,000. The amount of enzymes appears to be small in comparison to the venoms of terrestrial animals. A common characteristic of the toxic proteins of venomous fishes is their relative instability at room temperature or higher. This property is made use of during the initial treatment of envenomated persons. The first measure is to wash the wound with large quantities of water to extract as much as possible of the venom. After that, the injured part of the body is immersed in water as hot as the patient can tolerate for up to an hour, so the heat can denature the instable toxins.

Stingrays

Stingrays are found in all temperate and warm waters. With the exception of one family, which is confined to freshwater, rays are mostly marine animals, but some species may enter brackish water and even fresh water. They are primarily shallow-water inhabitants along coastal areas. Stingrays have always been recognized as being venomous and dangerous to man. In the ancient zoological literature they are referred to as "devilfish" and "demons of the sea."

The European stingray (*Dasyatis pastinaca*) inhabits the northeastern Atlantic Ocean, Mediterranean Sea, and Indian Ocean, and is a common species (Figure 34). The South American freshwater stingray (*Potamotrygon motoro*) is found in the fresh water rivers of Paraguay, and in the Amazon River. This species is extremely dangerous and its sting is reputed to cause intense pain. The round stingray (*Urolophus halleri*), found along the coast of California, is also dangerous to man.

Stingrays usually possess one, or occasionally two or more, barbed spines situated near the base of the long tail. They have a habit of burying themselves in the sand or mud in shallow waters. When someone treads on them they respond by a sudden vertical thrust of the tail, and the spine may penetrate through the skin of a

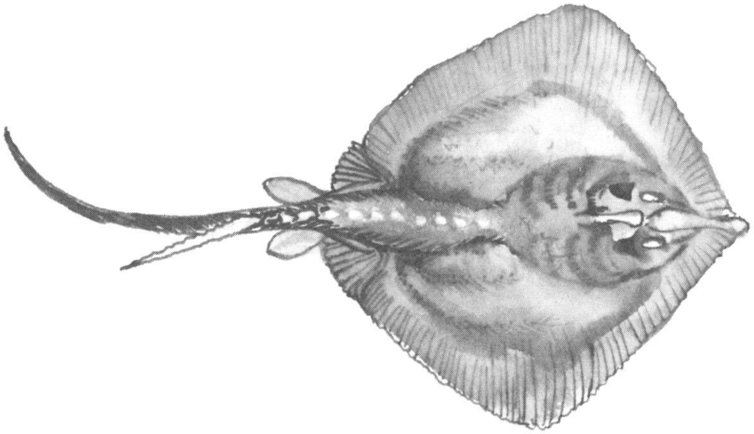

Figure 34. The European stingray, Dasyatis pastinaca, may be one meter in length.

human foot or leg. The spine does not have venom glands but toxins are produced by specialized cells within a sheath which surrounds the spine. When these cells are ruptured, the toxin is released into the wound. The venom is not very well characterized and it has been suggested that one potent toxin is a protein of intermediate size. Furthermore, the venom contains serotonin and two enzymes, 5-nucleotidase and phosphodiesterase. In humans, the venom causes extreme pain followed by several effects, including weakness, tremors, vomiting, fever, cardiovascular disturbances and more rarely, death. In contrast to many other potent venoms, this venom does not affect neuromuscular transmission. At intravenous injection in mice, the LD_{50} value of the toxic protein fraction of the venom is about 2.9 mg protein per kg body weight.

Weevers

Weevers are perhaps the most venomous fish found in temperate marine waters. They have a wide distribution, from the North

Figure 35. Lesser Weever, Trachinus vipera, becomes 10–15 cm in length.

Sea down to West Africa. The Great Weever (*Trachinus draco*) inhabits deeper waters than the Lesser Weever (*Trachinus vipera*), the Latin name of which refers to adder. The Great Weever occurs from Norway down to the coast of North Africa. The Lesser Weever inhabits the North Sea, southward along the coast of Europe, and the Mediterranean Sea (Figure 35). During spawning in the summertime, the Lesser Weevers gather in shallow waters. At English seaside resorts one can then see notices announcing "Warning Weevers."

The Lesser Weever is a short, stout-bodied fish up to 15 cm long. It has five to seven venomous spines in its dorsal fin and a single one on each gill. The spines are connected to venom producing cells contained within the integumentary sheath surrounding each spine. The weever buries itself in the sand with only the black dorsal fin protruding above the surface. There it waits for prey, such as shrimps or crabs, to pass by. Bathers may tread on the animal, sustaining a sting from one or more of the spines. It has been suggested that the sole, which buries in the sand, by only exposing one of its black fins gets some protection by mimicing the weever. The Great Weever is responsible for stings which fishermen may get. In France, weever is regarded as a delicacy and is caught on a commercial scale, but in some French cities the law requires that the spines must be removed before sale, as they can still inflict a painful sting.

The weever venom contains serotonin and some low molecular weight factor that releases histamine, which may explain the severe burning pain typical of weever stings. The lethal components are high molecular weight proteins. Physiological effects are similar to those caused by stingrays, and the toxicity is also comparable.

Stargazers

About 30 different species of stargazers are known. They mainly inhabit warm and tropical waters. One species, the common European stargazer (*Uranoscopus scaber*), is found in the eastern Atlantic and the Mediterranean. Stargazers are rather small (15 - 30 cm long), bottom-dwelling fishes. Like weevers, they spend a lot of their time buried in the sand where only the eyes and a portion of the mouth peep out. They also possess a very peculiar structure. It is a long slender worm-like process extending from the mouth region. This structure is used as bait to attract presumptive prey.

The venom apparatus consists of two sharp and conical shoulder spines, each of which protrudes through an enveloping integumentary sheath. Venom glands are associated with these spines, which have a double groove through which the venom flows into the wound of the victim. Stargazers may be fatal to man, but are not considered as dangerous as weevers.

Stonefish

There are a number of species of Stonefish (*Synanceja*), which are considered the most dangerous of phanerotoxic fishes. Stone-

Figure 36. Stonefish, Synanceja verrucosa, is venomous and difficult to detect by its camouflage color. It becomes 15–20 cm in length.

fish are members of the family Scorpaenidae, the scorpion fishes, most of which are venomous. One important and common species is *S. verrucosa*. It is found in shallow waters and is widely distributed in large parts of the Indo-Pacific region. This fish can attain a length of up to 15 to 20 cm (Figure 36) and in contrast to most venomous fishes, are camouflaged colored. They often lie on the bottom, partially buried in the sand and covered with slime and algae, resembling a large clump of mud. Because they are not easily seen, they are particularly dangerous for persons diving or wading in waters where Stonefish are common. The venom apparatus of this fish differs from that of other venomous fishes. It has 13 sharp and short spines along the back, 3 anal spines and 2 pelvic spines. The spines are associated with unusually large venom glands, which are covered by a very thick layer of warty skin. The spines are retractile and can rapidly be protruded to penetrate deep into the victim. Stings are extremely painful, and the pain may continue for a number of days. The pain may be so excruciating that the victim loses consciousness. The physiologi cal effects are extensive and include both cardiovascular and neuromuscular disorders. The toxic factors are supposed to be high molecular proteins, whereas low molecular compounds cause the intense pain. Stonefish account for a number of human fatalities. Fortunately, commercial antivenin is now available against its venom.

Zebrafish

The Zebrafish (*Pterois volitans*) is a beautifully colored fish, sometimes called the Butterfly Cod, which belongs to the scorpionfishes (Figure 37). It is widely distributed throughout warm and tropical seas and is common around rocks and coral reefs. Fortunately these fishes, in contrast to Stonefishes, are easily detected because of their brilliant colors and long and gracile fins. The venom apparatus consists of long and slender spines and associated venom glands. When they are provoked, the stinging spines project forward for defense. The Zebrafish are popular as aquarium fish and because of this, envenomations from their stinging spines are not uncommon. Like most other venomous fishes, they cause very severe pain. Other symptoms resemble those caused by Stonefish. Some deaths have been attributed to Zebrafish which, however, are not considered as dangerous Stonefish.

Moses Sole

Soles are probably more often associated with gastronomical joys than with poisons. Still, there are some very venomous sole species. The most well known of them is the Red Sea flatfish also called Moses sole, *Pardachirus marmoratus*. This fish is also considered a delicacy. Before it is served, it must, of course, be thoroughly

Figure 37. Zebrafish, Pterois volitans, has a beautiful and strange exterior with long venomous spines.

cooked to render the toxins harmless. This fish attains an average length of 20 to 30 cm. When Moses parted the Red Sea, a fish happened to be in the middle. The fish was split into two halves, through which the first specimens of a new species appeared, which consequently was called the Moses sole by the Israelites.

The venom apparatus of this sole is unique among fishes. A milky venom is produced by about 240 glands located along the dorsal and anal fins. The venom contains various very potent toxins. Minute amounts of the venom can be secreted through tiny pores into the immediate surroundings. It has been suggested that the Moses sole constantly exudes very small doses of the venom, and in this way forms a protective barrier against a hostile environment. It appears to be a very efficient shield since even voracious sharks avoid this fish. In one experiment, the venom was diluted 5,000 times and within a couple of minutes still killed all smaller fish which were exposed. The possibility of using one or more components of the venom as shark repellent for swimmers and divers has been considered.

One group of venom components is made up of steriod monoglycosides, the so-called mosesins. These substances in part exert the toxic and shark repellent activity of the venom. Other toxins are peptides, called pardaxins. Each pardaxin consists of 33 amino acids and, strange as it may seem, show marked physical and pharmacological similarities to melittin, the major toxin of the honeybee. In spite of this, the amino acid sequences of pardaxin and melittin are different. Pardaxins also contribute to the fish's defense against sharks and other predators. Pardaxins, like melittin, show cytolytic effects and, among other things, destroy red blood cells. In the venom, there is also a low concentration of a factor which can inhibit the cytolytic effects. It is possible that this factor can explain the sole's resistance to its own deadly venom. When isolated, the inhibitor also prevents the cytolytic effects of venoms from bees and certain snakes and therefore it also attracts much interest because of its conceivable applications in medicine.

17

Cryptotoxic Fishes

About 700 different cryptotoxic fish species are known. Most of them are stationary and are found in tropical waters, around islands, and along coral reefs. Some have tissues which are constantly toxic, whereas other species are only toxic during certain periods, or in certain regions. The poison can be diffusely spread in the whole organism, or in some cryptoxic fishes, be restricted to special tissues or organs. In most cases, the poison is believed to be due to the feeding habits of the fish, for instance, the ingestion of toxic plants and dinoflagellates by plant-eating fishes (herbivores). Carnivorous fishes can then accumulate the toxins by eating the poisonous herbivores. The toxic components do not affect the fish, but may be lethal to consumers higher up in the food-chain, for instance, to man or other fish-eating mammals. Intoxication caused by ingestion of a cryptotoxic fish is often called ichthyosarcotoxism.

Miscellaneous Poisonous Fishes

Eating the flesh and, in particular, the liver of some shark species has caused a number of envenomations and even deaths.

Most illnesses have been caused by tropical species. However, the flesh of the Greenland shark has been implicated in intoxications in both humans and sled dogs. These toxic factors are not known. It is possible that the toxin is an acetylcholinesterase inhibitor related to ciguatoxin (see below).

Lampreys and hagfish (cyclostomes) have been reported to cause poisoning. The toxicity could be due to bacterial contamination of these fishes, and the slime of the hagfish is suspected to be involved in some cases of poisoning. However, the envenomations do not seem to be very severe and the nature of toxic factors remains to be identified.

Mackerel, tuna, bonito and some other species belonging to the family Scombridae are generally edible and important food for many of the world's population. The flesh of certain species—particularly of those in tropical areas—is very susceptible and rapidly becomes tainted after the fishes are caught. The reason for this is that these fishes normally contain rather high amounts of histidine. When the fish is dead, histidine is transformed by special bacterial enzymes to histamine and related substances, which can cause severe allergy and other poisoning symptoms. For some unknown reasons, these chemical changes are much more likely to occur in scombroids than in other fish species.

Fishes Containing Ciguatoxin

A most problematic type of poisoning is produced by a large variety of fishes; this is responsible for several human intoxications and even deaths. It is called ciguatera and is caused by ciguatoxin. More than 440 marine species are regarded to be potentially ciguatoxic, and ciguatoxin may also appear in certain marine invertebrates, e.g., echinoderms, molluscs, and arthropods. The majority of toxic fishes are tropical reef or shore fishes and usually bottom-dwellers. Examples of ciguatera-producing fishes are surgeonfishes, sea basses, mullets, trunkfish, herrings, and eels. Since these fishes are valuable food fishes, ciguatera poisoning is both a common and an unpredictable type of poisoning. It especially is a serious problem in certain tropical areas, such as the central and southern Pacific Ocean, and the West Indies. Ciguatoxin is very treacherous since it does not affect the taste of the fish and is potent at very low concentrations. Unfortunately the toxicity of fishes in some tropical areas is also unpredictable because it can appear very quickly, only to disappear some years later. Poisoning causes muscle tiredness, decreased coordination of muscle movements, vomiting, diarrhea, and increased parasympathetic activity.

Ciguatera poisoning is associated with the feeding habits of the fish (Figure 38). Many fishermen in the tropical archipelago believe that the fish becomes poisonous by eating various corals and algae. However, it now seems established that ciguatoxin is elaborated by a dinoflagellate discovered in French Polynesia and identified as *Gambierdiscus toxicus*. By ingestion of such dinoflagellates, the toxin is transferred to herbivorous fish, which in its turn is eaten by man. The poisonous herbivorous fish may of course first be eaten by a carnivorous fish, which subsequently is eaten by man. The toxin may also originate in various marine invertebrates and be passed along a more complex food chain to man.

Although ciguatoxin has been subjected to extensive chemical and pharmacological investigations, its mechanisms of action are not fully understood. This can in part be explained by difficulties encountered in extracting it in quantities enough for chemical analyses. As an example of the low recovery one may mention a recent investigation in which it was possible to extract only about 1.3 mg ciguatoxin from 50 kg of moray viscera. Ciguatoxin has a molecular weight of 1,110. It shows lipid solubility and good temperature stability. It is extremely potent with an LD_{50} value of 0.45 μg/kg at intraperitoneal injection in mice. It has been observed to inhibit the propagation of action potentials in nerves, and also exerts cholinergic effects on nerve terminals, which may be due to its inhibition of acetylcholinesterase, an enzyme which breaks down acetylcholine.

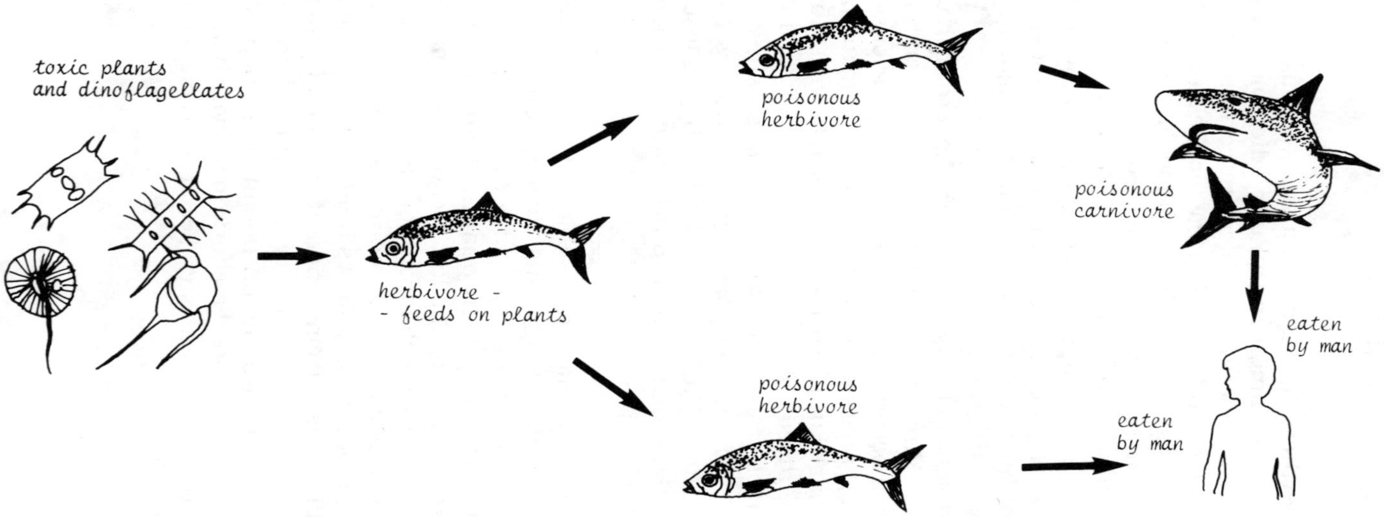

Figure 38. Possible chain for transfer of toxins of dinoflagellates.

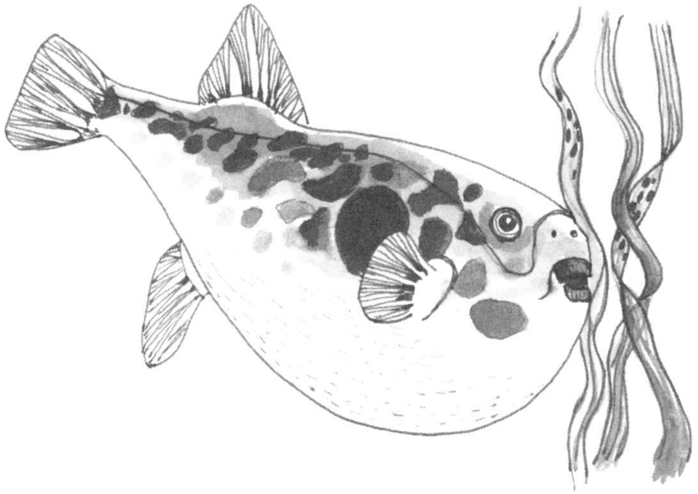

Figure 39. Puffer.

Fishes Containing Tetrodotoxin (TTX)

Tetrodotoxin is one of the most potent animal toxins we know. Since it is also found in a special fish which is regarded as a delicacy in the Far East, it is no wonder that there is an extensive literature about tetrodotoxin, the most studied and well known of the animal toxins. The fishes are called blowfish or more often puffers (Figure 39). The name tetrodotoxin (TTX) is derived from their Latin family name, Tetraodontidae. TTX may also be found in other kind of fishes, for instance in trunkfishes, filefishes, sunfishes, and triggerfishes. About 75 different fish species are known to contain TTX. However, the puffers are best known and also economically most important.

Puffers are most numerous in the tropics but several species also inhabit temperate regions, for example, there are some poisonous species in North American waters. In *The Log From The Sea Of Cortez* Steinbeck describes how they asked a boy in Baja California if they could buy a puffer fish he was carrying. The boy

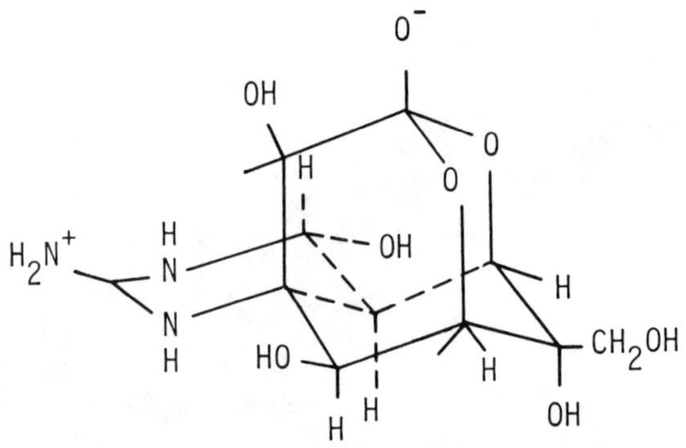

Figure 40. Tetrodotoxin (TTX).

refused saying that a man had commissioned him to get this fish and he was to receive ten cetavos for it because the man wanted to poison a cat.

The names of these fishes (Blowfish, Puffer) refer to their characteristic spherical shape which they may attain by filling a bladder, connected to the digestive system, with water or air. This shape, in combination with sharp teeth, provides efficient protection against enemies. The remarkable appearance of puffers has always fascinated man. Puffers have been depicted in Egyptian graves from 2700 B.C. The toxicity of puffers was already stated by the legendary emperor of China, Shun Nung (2838-2698 B.C.). He personally tasted 365 drugs while compiling a pharmacopeia, but in spite of his risky occupation enjoyed a long and happy life.

The chemical structure of TTX is unique, and is not similar to any other known molecule (Figure 40). It is a small, ring-shaped spherical structure ($C_{11}H_{17}N_3O_8$, mol. weight = 319). The highest concentrations of TTX in the fish are found in the liver, intestines, kidneys,and ovaries. TTX specifically blocks the voltage-sensitive Na^+ channels and therefore inhibits the inward Na^+ current (Figure 41). This toxin only exerts its action on the outside of the membrane and has no effect if it is introduced on the inside of an excitable cell. Thus, if poisoned with TTX, the electrical activity ceases in nerves and muscles. One great advantage of TTX for

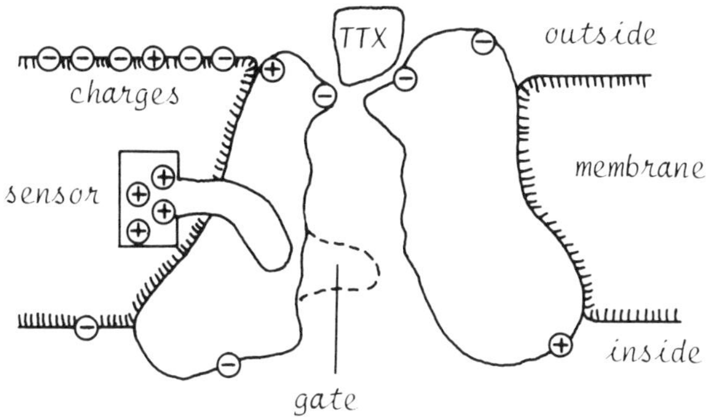

Figure 41. Mechanism of action of tetrodotoxin (TTX).

neurophysiological studies is that its action is highly specific. Among other things TTX has been used to calculate the molecular dimension of the Na^+-channel and for the analysis of synaptic transmission in the absence of nerve impulses. With the possible exception of saxitoxin (STX), tetrodotoxin is the most specific among the known blockers of the Na^+-channel. Interestingly, STX has recently been reported to be present in puffers. Its presence seems to be sporadic and the concentration is very low. We shall return to STX later.

Very little is known about the origin, synthesis, and physiological significance of TTX in fishes. TTX is not unique for these fishes. An increasing number of animals have been found to contain TTX, e.g., California newt (*Taricha torosa*), atelopid frogs (*Atelopus*), the Japanese ivory shell (*Babylonia japonica*), the trumpet shell (*Charonia sauliae*), a starfish (*Astropecten poycanthus*), and probably the blue-ringed octopuses (*Hapalochlaena maculosa* and *H. lunulata*). Even eggs of the latter species appear to contain TTX. Even if the biological role of TTX in these animals is unknown, it is possible that the toxin protects the animals against predators. Some observations indicate that the synthesis of TTX is dependent on symbiotic microorganisms or other factors in the natural environment, since TTX is absent in puffers brought up in fish cultures. It has been shown that TTX does not affect nerve fibers and muscles in the puffer fish and in the newt. This is not because of the immune system of the animals, but probably is due to the special nature of the Na^+ channel of animals in which TTX normally exists.

TTX is one of the most potent non-protein toxins known. The LD_{50} value is 10μg/kg at intraperitoneal injection in mice. It is 25 times more powerful than curare. About one mg would be enough to kill an adult man. The first symptom is a tingling or pricking feeling in the tongue, the lips, the fingers, and the toes, which within 10 to 45 minutes is followed by uncoordinated movements. After that, the sense of feeling is lost, followed by weakness, vomiting, and diarrhea. In severe cases, the affected person dies in respiratory paralysis. Several envenomations, sometimes lethal, occur each year in countries where puffers are eaten. In spite of its toxic effects, TTX is used in a very dilute form in Japan as an analgesic for neuralgia, arthritis and rheumatism.

In Haiti, some voodoo sorcerers are able to put their victims into a cataleptic state through the administration of a powder made of native plants and animals containing toxic substances. The animals include puffer fish, frog and toad species. It has been shown that the powder contains TTX, to which some of the dramatic physiological effects may be attributed.

In Japan, puffers are called fugu like the classical dish based on this fish. In Japan eating fugu has always been an almost sacred procedure, taking place in a devotional spirit. The obvious health risks associated with this habit appear to count for little compared with the possibility of enjoying this gastronomical highlight. Somebody has compared a fugu dinner with a gastronomical version of Russian roulette. About 60% of fugu poisonings prove fatal. Until 1966, about 100 persons died each year in Japan from eating fugu. Since then the mortality rate has declined but the following verse is still unfortunately relevant:

> "Last night he and I ate fugu;
> Today, I help carry his coffin."

To obtain a license to serve fugu, the cook must have taken a special course, which includes intense practical training and a written examination. The preparation of the meal is a work of art. There are 30 steps prescribed by law for preparing fugu, which take about 20 minutes for an experienced cook to execute. Among gourmets, or rather dare-devils, the liver has a special attraction, for it is one of the most poisonous parts of the fish and is not allowed to be served. A fugu is based on the raw flesh which usually is cut in thin, transparent slices and arranged in exquisite patterns forming beautiful birds or flowers. Eating fugu is regarded as a culinary delight but requires a bulging wallet. It is one of the most expensive meals in Japan. The chefs buy the puffers at special markets where big money is converted to a deadly poisonous, mysterious, exotic, and worshipped fish.

18

Animals Containing Saxitoxin (STX)

Another extremely poisonous toxin is saxitoxin (STX) (Figure 42). Like TTX it specifically blocks the voltage-sensitive Na^+-channels and therefore inhibits nerve impulses. It is regarded to be as specific and potent as TTX.

STX originates from several dinoflagellate plankton of the genus Gonyaulax and in particular from one species, *Gonyaulax catenella*. At times there is a local growth or bloom of this and other dinoflagellates producing a discoloration of the surface water, commonly referred to as "red tides," which cause a mass death of fishes and other marine organisms. Besides STX, the dinoflagellates synthesize a red pigment related to carotene, which gives the bloom its characteristic color.

STX at very low concentrations has been reported to occur in puffers and in some other fish species. However, poisoning through fish food is probably rare. On the other hand this and other toxins may accumulate in shellfish and other invertebrates (e.g., some crab species) through ingestion of planktons. If such animals are subsequently eaten by man it may lead to poisoning, the so-called paralytic shellfish poisoning. The toxin responsible is often called paralytic shellfish poison (PSP). This kind of poisoning is a serious economic and public health problem in some

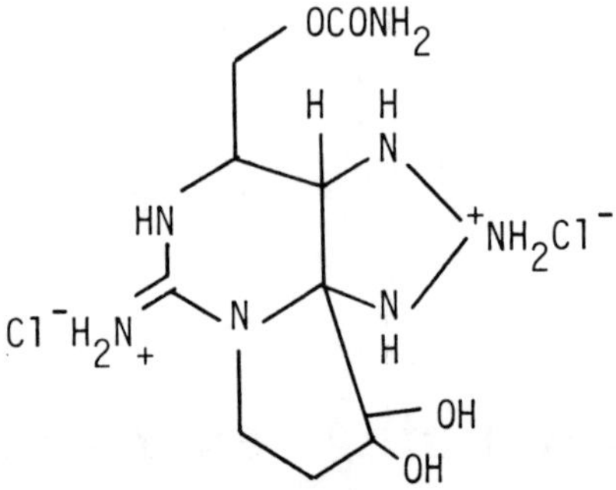

Figure 42. Saxitoxin originates from dinoflagellates.

areas of the world. Poisoning involves a general feeling of sickness, fever, diarrhea, and in severe cases, neurotoxic symptoms, and paralysis. This more difficult form of poisoning has been described in the literature already during the eighteenth century and has caused several human deaths. Milder forms of poisoning are more common, and may be caused by toxins of other algae; as a result, periodically there have been bans on the sale of various edible mussels.

Horsehoe crabs are unique marine animals (Figure 43). They are related to spiders and may be regarded as living fossils, which for 300 million years have kept their typical appearance. The rounded plated back is equipped with a characteristic motile dagger-shaped tail. There are only three genera (*Limulus*, *Carcinoscorpius* and *Tachypleus*) and four species living today: *L. polyhemus*, *C. rotundicauda*, *T. gigas* and *T. tridentatus*. The last three species are found extending from Japan south to the Indo-Pacific areas. Asian horseshoe crabs are delicacies appreciated in many parts of Southeast Asia; the unlaid eggs are especially popular.

Horseshoe crabs, in particular *C. rotundicauda*, appear to be poisonous to eat at certain seasons of the year. There is also a large individual variation in the toxicity of the horseshoe crabs. There are several reports of poisoning resulting from ingestion of

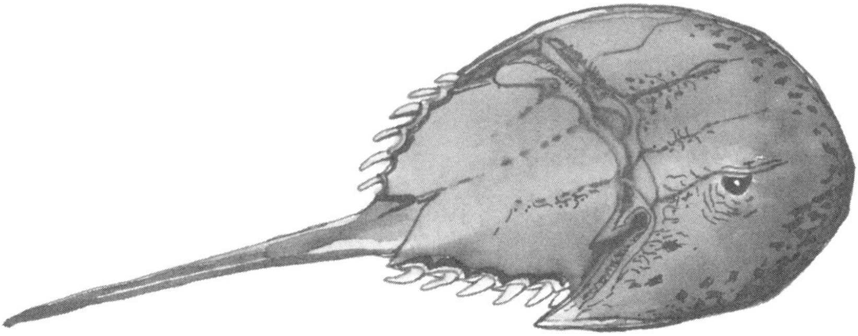

Figure 43. Horseshoe crab.

C. rotundicauda, which in some cases proved fatal. The victims show severe neurological symptoms. It now appears established that the toxin is identical to saxitoxin, which the horseshoe crabs accumulate through their food.

19

Amphibians (Amphibia)

The science of the biology of amphibians and reptiles is classified as herpetology. These animal classes have been extensively studied and have attracted special attention from toxinological point of view. One reason for this is that they may cause the most important and common poisonings of animal origin. Another reason is that several toxins in their venoms have been used as experimental tools, and have contributed to significant progress in physiology.

There are today about 3,000–4,000 amphibians, which are divided into frogs and toads (Anura), salamanders and newts (Urodela), and wormlike caecilians (Gymnophiona). Amphibians are poikilothermic vertebrates. Most are equipped with a soft and moist skin, which is of great importance for respiration and the balance of water and salts. It is also common that the skin is rich in pigment cells, which have a role in camouflage. Furthermore, several species use venomous secretions for defense.

Dart-Poison Frogs (Dendrobatidae)

Dart-poison frogs represent, by their spectacular colors, noisy behavior, remarkable way of nursing the tadpoles, and not least their extreme venomousness, a very fascinating group of animals. They are found in varying habitats in South America and southern Central America. Some species reside in the crowns of the rain forests, to which the Latin family name, Dendrobatidae, refers. Other species prefer low-lying areas or pools of water. In contrast to most other frog species, dart-poison frogs are active only during the day. They are unconcerned about their small size (2–6 cm long) and are responsible for an impressive part of the noise which can be heard during the day in the rain forests.

Dart-poison frogs comprise about 130 species, divided into 4 genera; *Atopophrynus, Colostethus, Dendrobates* and *Phyllobates.* The about 50 Dendrobates and 5 Phyllobates species all show a bright and conspicuous coloration. In contrast to many other amphibians, which are camouflage colored, these small frogs send out a warning. We are deadly poisonous, keep away! When danger threatens, the frogs secrete a complicated solution of toxic components from miniature glands in the skin. Animals which eat these frogs may die, but if they survive, they do not repeat the blunder. Very few animals are completely protected from predators and even the dart-poison frogs are prey for certain snakes and large spiders. It has been suggested that the poisonous secretion also contains components which protect the moist frog skin from infections by microorganisms, for which the rain forest climate provides excellent conditions. These secretions have been shown to inhibit the growth of bacteria and fungi at very low concentrations. Frogs other than the dart-poison frogs, for instance those within the genera *Phyllomedusa, Rana,* and *Xenopus,* are also known to secrete components which play a defensive role but also show antimicrobial activity. Recently, peptides with broad spectrum antimicrobial activity have been found in high concentrations in the skin of *Xenopus.* They appear to be discharged on injury to the skin, and also if stressed by predators. The peptides are clearly noxious to predators and cause remarkable effects. For example, *Xenopus* secretions cause a garter snake to yawn by inhibition of the mouth closure mechanism.

Defense by mimicry, which is so common among insects, also exists among frogs. Some nonpoisonous frogs have a coloration showing deceptive similarity to that of the dart-poison frogs, which deludes presumptive predators.

Three Phyllobates species are the most toxic among the dart-poison frogs. They are found along the coast of Colombia and along the western slopes of the Andes. East of the Andes the Indians use curare, prepared from various species of plant, to poison the arrows. However, in the western Colombia the Indians impregnate their arrows and darts with the skin secretions of Phyllobates. Due to access to modern firearms, the use of poisoned arrows is fortunately becoming rarer. Captain Cochrane, who made expeditions to Colombia during the early nineteenth century, reported how the poor frogs were milked for their toxic secretions by the Indians. A pointed piece of wood was passed through the mouth and out at one of the legs of the tortured frog. This made the frog perspire until it became covered with white froth, which the Indians collected and used to cover the points of their arrows. One frog provided ammunition for about 50 arrows. The Swedish anthropologist Henrik Wassén made expeditions to Colombia in 1934 and in 1955. He writes that two Chocó Indian tribes along the San Juan River used venom from *Phyllobates bicolor* and *Phyllobates aurotaenia* for preparing their arrows.

The active components of the venomous skin secretions of the dart-poison frogs are alcaloids; nitrogen containing aromatic molecules. Toxic alcaloids are otherwise dominating among plants but more unusual in animal venoms. However, dart-poison frogs elaborate a remarkable diversity of biologically unique alcaloids, of which about 200 new alcaloids have been identified. Of these, at least a dozen are potent toxins, which are used for defense against predators.

In venoms of Phyllobates species the dominating toxins are the so-called batrachotoxins (batrachos is Greek and means frog). These complex alcaloids show structural similarities to steroids, and represent some of the most toxic non-protein substances known (Figure 44). Batrachotoxins specifically block the inactivation of the voltage regulated Na^+ channels in nerve and muscle cells, which causes a massive inflow of Na^+. The cells become irreversibly depolarized, which among other things produces

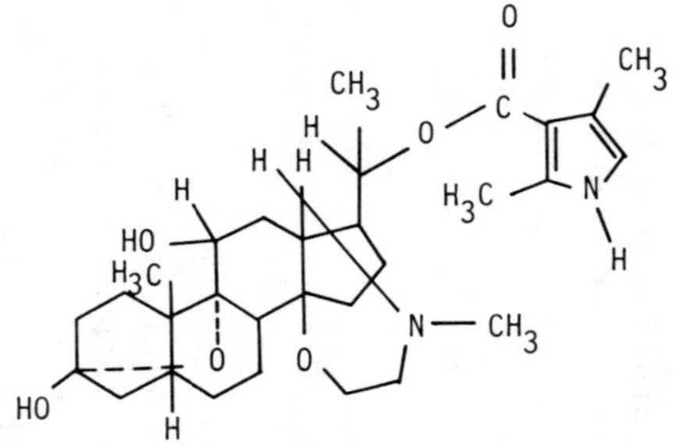

Batrachotoxin

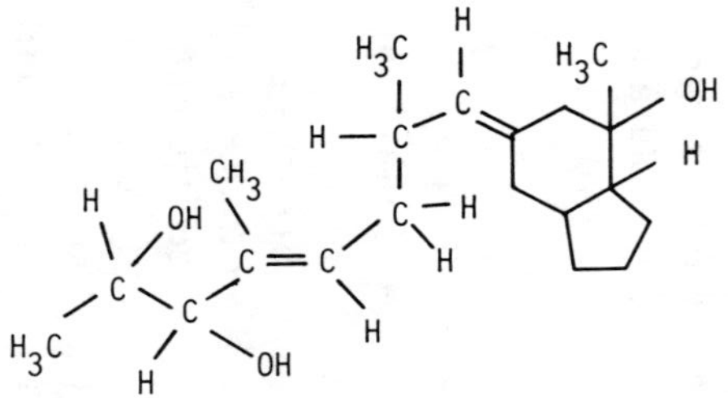

Pumiliotoxin B

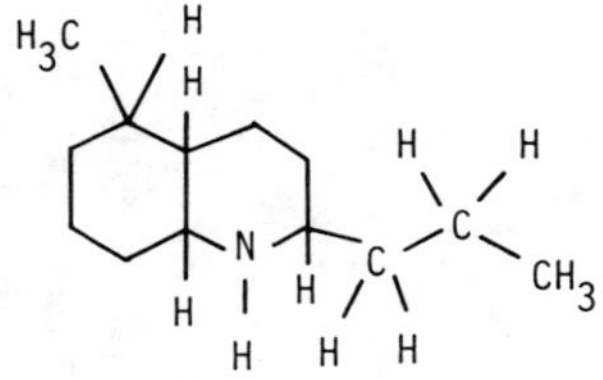

Pumiliotoxin C

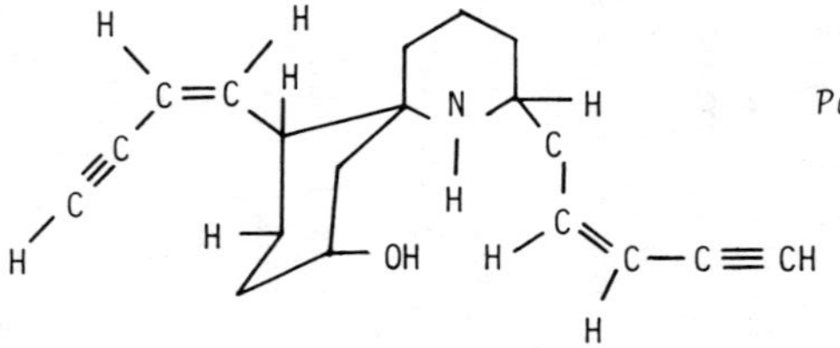

Histrionicotoxin

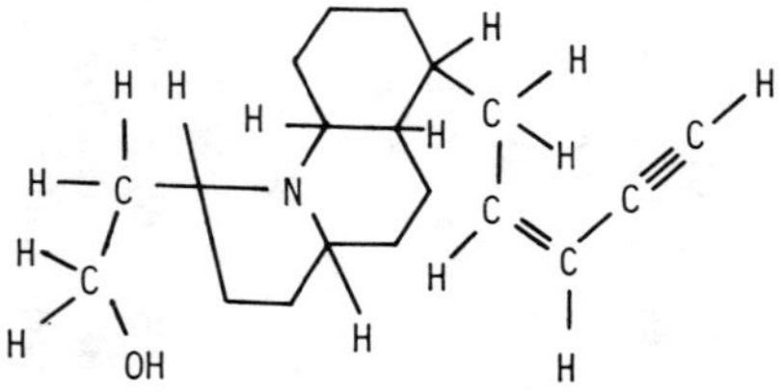

Gephyrotoxin

Figure 44. Toxins from dart-poison frogs.

Table 3. Animal toxins affecting voltage regulated Na^+-channels.

Species	Toxin	Mechanism
Anemonia sulcata (sea anemone)	ATX II, basic protein. Mw = 4770	Causes slow and incomplete inactivation of the Na^+-channel. The duration of the action potential is increased.
Puffers, California newts, Trumpet shells, Japanese ivory shells, Atelopid frogs,	TTX, cyclic structure. Mw = 391	Inhibits activation of the Na^+-channel.
Scorpions	Basic proteins. Mw = 8000	3 different types: 1, delays in activation of the Na^+-channel. 2, reduces the threshold for activation of the Na^+-channel. 3, reduces Na^+-and K^+-currents with no changes in the kinetic of activation and inactivation.
Dart-poison frogs	Batrachotoxin, complex alcaloid	Inhibits inactivation of the Na^+-channel. Irreversible depolarization

heart arrhytmia and respiratory failure, and finally cardiac insufficiency. A lethal subcutaneous dose for a mouse is about 100 ng (10^{-9} g). It has been calculated that about 0.2 mg would be enough to kill a man. We have previously acquainted ourselves with toxins of other animals which affect the cellular Na^+-channel (Table 3). Like batrachotoxin, these other toxins are very potent, which may illustrate the strategic importance of the Na^+ channel for the survival of animals.

The remaining frog alcaloids are chemically somewhat simpler and also less toxic than the batrachotoxins. Pumiliotoxin B is a unique indolizidine alcaloid, which was first isolated from *Dendrobates pumilio* (Figure 44). It affects the transport of intracellular Ca^{2+}. When a nerve signal reaches an innervated muscle, pumiliotoxin B causes the release of Ca^{2+} from the intracellular muscle reservoirs, which in turn elicits a contraction. Pumiliotoxin B stimulates the release of Ca^{2+} and inhibits restoration of the normal Ca^{2+} depots. In this way, pumiliotoxin B increases and prolongates muscle contractions which can lead to spasms and convulsions.

Figure 45. The Aga toad, Bufo marinus, belongs to the biggest and most venomous toads. It may be over 20 cm in length.

The possibility of using pumiliotoxin B in low concentrations or in modified form as a heart medicine has been considered.

From the skin secretion of *Dendrobates histrionicus* other toxins have been isolated, histrionicotoxins, which are so-called spiroperidine alcaloids. These toxins are not particularly potent, but are still interesting pharmacological tools with properties similar to those of curare. They inhibit the cholinergic nicotinic muscle receptor. Gephyrotoxins, still another group of frog alcaloids, also affect cholinergic transmission and extend the arsenal of useful instruments for future research in the physiology of nerves and muscles.

Toads (Bufonidae)

There are about 150 toad species (Bufonidae) distributed over large areas of the world. All toad species secrete toxic substances in the skin. These venoms are produced by special parotid glands

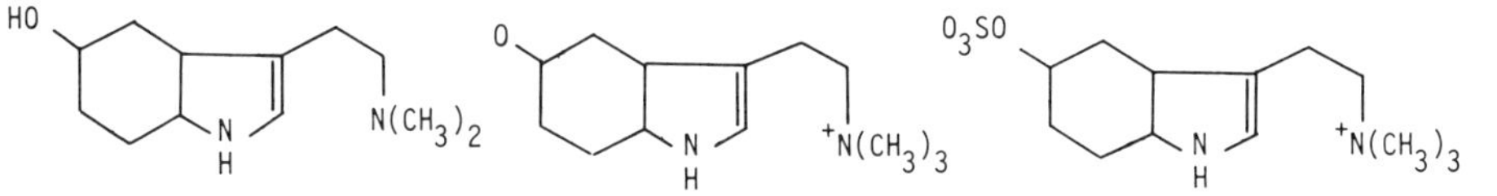

Bufotenin Bufotenidin Bufoviridin

Figure 46. Hallucinogenic compounds from toad skin.

and/or miniature glands scattered in the skin. Tadpoles of certain species also secrete a venomous substance. In comparison to that of the dart-poison frogs this secretion is relatively harmless but effectively discourages most predators. It is also believed to contain components which inhibit microbial skin infections.

Among the most toxic toads are the Colorado River toad (*Bufo alvarius*) and the Marine River toad or the Aga toad (*Bufo marinus*) (Figure 45). In the United States there are annually several cases of toad poisoning, usually dogs and cats, which may have played with the animals and bit or mouthed the toads. It has been estimated that more than 50 dogs a year die from toad poisoning in Hawaii. In northern Europe, frogs comprise a considerable part of the diet of the otter (*Lutra lutra*), which avoids toads because of their poisonous skin secretions. There are of course also some enemies of toads. For instance, hedgehogs and polecats appear to be rather resistant to toad venoms, but not completely, since they always leave the skin after a toad meal.

The chemical composition of the toad secretions is characterized by great variety. Besides biogenic amines such as adrenaline, noradrenaline and dopamine, there are unusual indoalkylamines, e.g., bufotenin, bufotenidin, and bufoviridin (Figure 46). The latter show chemical similarities to 5-hydroxytryptamine (serotonin), a nervous system transmitter, and to the hallucinogenic compound, LSD. In man bufotenin also produces symptoms similar to those of LSD. When the girl in the fairy-tale kissed a frog—or was it a toad—it was changed into a prince. Could it possibly be explained by hallucinations?

In addition, all analyzed toad skin secretions contain the so-called bufodienolide toxins (Figure 47), which are also found spo-

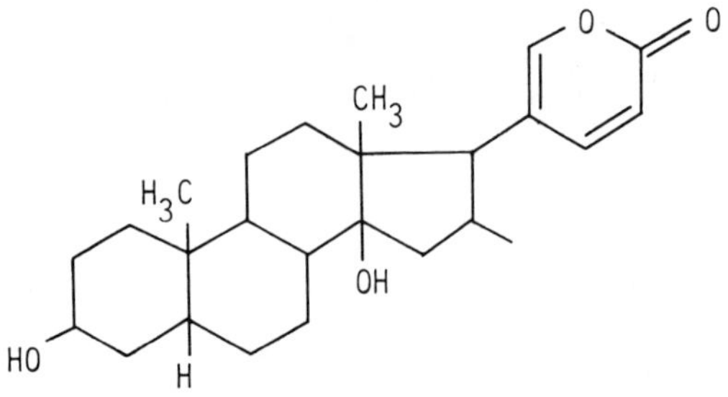

Figure 47. A toad toxin, bufotalin, belongs to the bufodienolides.

radically in many frog species, but at a much lower concentration. The toxic effects of bufodienolides are due to their potent inhibitory effects on the membrane enzyme, Na^+,K^+-ATPase, which is responsible for pumping Na^+ into and K^+ out of the cells. We have previously met such inhibitors in the discussion of the defense mechanisms of echinoderms (steroidal and triterpene glycosides). In the vegetal world they correspond to the cardenolides, among which digitoxin and ouabain are the more well known. In toads, like in amphibians in general, the skin is the most important tissue for the regulation of the salt and water balance of the body. In the amphibian skin there is also a remarkably high Na^+, K^+-ATPase activity. It has been speculated that the bufodienolides have been developed from some factor which normally should regulate this enzyme activity.

Salamanders and Newts (Urodela)

Research during recent years has shown that several different salamander and newt species contain very potent toxins, which protect against predators. The extremely potent substance tetrodotoxin (TTX), which we have earlier discussed in connection with the poisonous fishes (puffers), has been found in at least two salamander species: the red eft (*Notophthalmus viridescens*), found in eastern North America, and the California newt (*Taricha torosa*). Envenomation of humans by these animals is rare. However, poisoning of domestic animals may occur.

Among the lungless salamanders (Plethodontidae), there are two American species which are very popular for collectors, the red salamander (*Pseudotriton ruber*), and the mountain salamander (*Pseudotriton montanus*). It has previously been considered that the red salamander is a nonpoisonous animal, which by a similar color and appearance mimics the very toxic red eft, thus obtaining some protection to predators. However, it has now been shown that the red salamander is far from harmless, having a very noxious skin secretion comparable in toxicity to TTX. This toxin is called pseudotritontoxin. In contrast to TTX and typical amphibian toxins, which are of low molecular weight, pseudotritontoxin is a high molecular weight compound (mol. wt. $\geq$ 200,000). Its properties and mechanisms of action are still poorly known.

Other venomous salamanders may be mentioned, for instance the European Fire salamander (*Salamandra salamandra*), whose bright colors announce the presence of venomous skin secretions. The venom, often called salamandrin, is a milky liquid produced by glands situated behind the ears. The secretion is very irritating and very painful for the eyes and mucous membranes. It provides good protection against most animals, except for humans, who wrongly believe that the animal is very dangerous and unfortunately often kill them. Much remains to be learned about the toxinological aspects of the physiology of salamanders and newts.

20

Reptiles (Reptilia)

Reptiles are poikilothermic animals which use lungs for breathing. The skin bears horny scales, which may or may not be reinforced by bone. The young emerge from eggs laid on land. Birds and mammals are descended from reptiles, which are a very old group. The "Age of Reptiles" was during the Mesozoic period beginning about 230 million years ago. At the end of the Cretaceous period, the majority of the species disappeared, e.g., the dinosaurs, for reasons which are still unknown. The living representatives comprise about 6,500 species divided into several different orders, of which the most interesting one from toxinological point of view is the Squamata, since this order includes the suborders lizards (Sauria) and snakes (Serpentes). They represent 95% of all reptile species. Most lizards and snakes are quite different from each other, and it may be surprising that they are grouped together. However, a close examination shows that they have several characteristics in common and almost every snake feature can be matched in some lizard. A few lizard and a large number of snake species have developed an advanced and specialized venom apparatus, consisting of salivary glands transformed to venom glands, and modified teeth for injection of the venom.

Venomous Lizards (Helodermatidae)

Nearly 3,800 species of lizards (Sauria) have been described. Of them, only two are venomous and have developed a special venom apparatus. The two species, the Gila monster (*Heloderma suspectum*) and the Mexican beaded lizard (*Heloderma horridum*) (Figure 48), make up their own family with just one genus (*Heloderma*). The brilliant colors of these lizards announce that they are venomous. The colors are accentuated by the arrangement of the scales, which give the impression that the body is covered with shining pearls. Full-grown Gila monsters and Mexican beaded lizards are heavy-bodied and may be up to 50 and 80 cm long, respectively. They are active mostly at night and at dusk. Even if their gait is rather slow and awkward, they are capable of climbing and have been observed in trees in search of birds' eggs. The Gila monster lives in the desert areas of the southwestern United States and northern Mexico, and the Mexican beaded lizard lives in western Mexico. They eat nestling birds and eggs of birds and reptiles, small rodents, young cottontail rabbits, and possibly lizards. Amphibians appear to be resistant to their venom. The lizards themselves seem to have few enemies other than humans.

In contrast to snakes, which have the venom glands located in the upper jaw, the venom glands of the lizards are located in the lower jaw, with four glands on each side. The venom is transferred to the grooved teeth by capillary action. The grooves are not connected to the glands, which makes the venom apparatus less efficient. When biting, the lizards hang on with a bulldog-like grip, but may shift the hold several times and chew slowly while retaining its hold. The chewing movements facilitate the flow of venom into the wound. The lizards may sometimes turn upside-down, perhaps to increase the flow of venom into the victim. The venom is very potent and there is no specific antiserum yet available. However, the lizards are slow and not very aggressive, and humans are therefore seldom bitten. Fatal poisoning of humans seems to be rare. Reports that 25% of persons bitten have died poisoned are likely to be exaggerations.

The venoms of the Gila monster and the Mexican beaded lizard are complex mixtures of similar composition. There is compara-

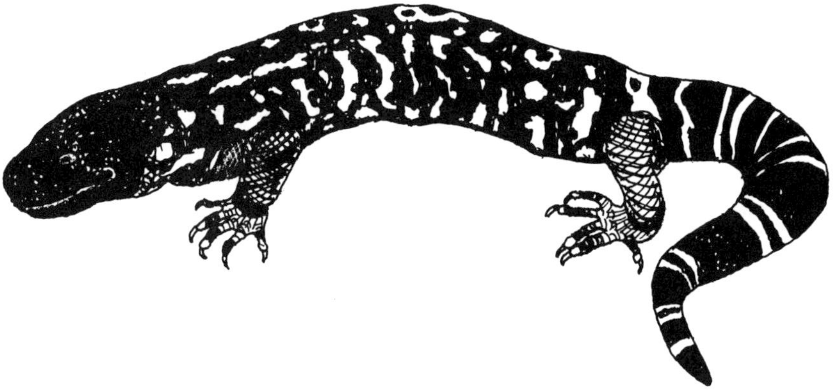

Figure 48. Mexican beaded lizard (Heloderma horridum) becomes about 80 cm in length.

tively little information on the chemistry of the venom and its mechanisms of action. It is considered to show some similarities to the venoms of elapids (e.g., cobra venom). The venom contains both neurotoxic and necrotic factors and as in several other venoms, a high hyaluronidase content is present. It serves to aid in the diffusion of toxins. In addition, the venom contains phospholipase A 2, which probably is responsible for the neurotoxic effects. Recently a lizard toxin causing partial paralysis of the rear limbs and the lowering of body temperature in experimental animals has been reported. Since the lizards are slow animals, it is possible that they have evolved specific toxins to immobilize their prey. A kinin-releasing factor has also been identified in Heloderma venoms, and this may contribute to the shock reactions of the victims.

Snakes (Serpentes)

Like lizards, snakes (Serpentes) are a very successful group of reptiles; about 2,300 species have been described. The snakes are closely related to the lizards and are believed to have originated from that group. In comparison to other reptilians, the snakes appear to be a rather young group, which has arisen during the Cretaceous period, about 125 million years ago. Despite their lack of limbs, snakes are represented in most parts of the world, except for the Arctic and Antarctic areas. Furthermore, venomous snakes are not present on several islands, for instance, in New Zealand and in Ireland.

The serpentine movements of snakes, their stretchable bodies, and in particular their venoms, have contributed to man's fascination and also fear of snakes from time immemorial. No other animals have been subjected to so much creation of myths as have snakes, which have a prominent place in folklore and legends. The snake has been hated and associated with the powers of evil, but has also been loved and worshipped. In ancient times snakes often symbolized the healing power of nature, and were thought to be the nearest to that from which human life and health came. In the major households a sacred snake was often kept and tended carefully. The ability to cast its skin was supposed to symbolize fertility. In ancient Egypt, the cobra was worshipped and its picture decorated the crowns of the emperor. In the old Greek world, the god of medicine had a stick entwined with a snake, a symbol still used by the guild of medicine and by pharmacists in many places of the world.

Snakes form a clearly distinguishable group of reptiles. However, their classification is controversial, and the number of families varies from 9 to 15, according to the textbook consulted. Of these, we shall only discuss families which have venomous species. They are the colubrids (Colubridae), the elapids (Elapidae), the sea snakes (Hydrophiidae), the true vipers (Viperidae) and the pit vipers (Crotalidae). This arrangement can probably be criticized but is practical when we discuss snake venoms. Some zoologists regard the sea snakes as a subfamily of the elapids, and the pit vipers as a subfamily of the true vipers. The colubrids represent the largest family (about 1,500 species) of which only a few species are venomous, whereas all the representatives of elapids (170 species), sea snakes (50 species), true vipers (40 species), and pit vipers (140 species) are venomous. It should also be pointed out that the number of species varies in different systematical compilations.

In the literature of animal toxinology, a dominant part is devoted to the study of venomous snakes. In spite of this, it is also true for the highly complex snake venoms that many exciting discoveries with respect to snake toxins can be expected in the future. A short review of the anatomy and physiology of snakes may be appropriate to better understand the mechanisms and function of the venom apparatus in the different species.

The snakes show many interesting peculiarities, one being the complete absence of limbs. Only in a few of the more primitive forms (e.g., boas and pythons) can vestiges of the hind limbs and their girdles be found. Despite the lack of limbs, many snakes can move rapidly, some of them much faster than a man can walk. Nearly all can swim, although several species avoid water.

One type of locomotion is produced by the lateral undulation of the body, which exerts pressure on surrounding objects and pushes the snake forward. The ventral scales of most species help to prevent slipping. Another type of movement is referred to as rectilinear locomotion or caterpillar action. The body is shifted forward within the skin and the snake flows along an almost straight course. Still another type of movement is achieved by throwing the body into S-shaped curves, followed by a quick straightening. It is therefore called the concertina kind of movement. A fourth type of locomotion, known as "side-winding," has been developed by several species of snakes living in desert areas. It is effected by throwing the body out in what approximates a helical roll, the snake progressing by walking, as it were, on the lower edge of each loop.

The eyelids of the snake have become fused to a transparent cover which is periodically shed with the rest of the skin. In most snakes, the lens is round and has a limited ability to change its shape. Instead, focusing is brought about by an increase in pressure in the chamber behind the lens so that the lens is forced forward. The snake probably cannot distinguish details in the surroundings, but effectively detects movements within range of its vision. With some exceptions, snakes generally have poor vision, and depend on other senses for the detection of prey or danger over longer distances.

Their sense of hearing is greatly modified, and external ears are lacking. Snakes are probably unable to detect airborne vibrations.

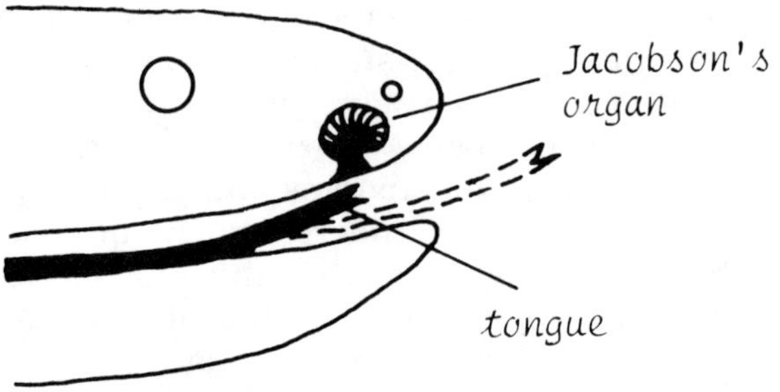

Figure 49. The Jacobson's organ is of great importance in smelling.

This means that no rattlesnake has ever heard its own rattle. However, the inner ear bones lie in contact with some bones at the rear of the jaw. Ground vibrations can therefore be transmitted through the jaw resting on the ground to the inner ear. This mechanism is likely to be of great value in detecting animals even at considerable distances.

The sense of smell is remarkably well developed among the snakes, and probably plays an important role in the detection of prey. Besides the nose proper, there is the Jacobson's organ, which appears to be of special importance in smelling (Figure 49). It consists of a pair of chambers above the front of the palate, each opening into the mouth by means of a duct. Odorous particles from the air or the ground are carried to the Jacobson's organ by the forked tongue tip. With this technique, the surroundings can be investigated and prey be traced. If a snake's tongue is removed, the animal is badly disabled.

Snakes also have very special receptors, heat sensitive receptors, which are highly specialized in pit vipers (Figure 50). The pits are situated between the eyes and the nostrils, and contain thermal receptors, which are capable of detecting very small changes in heat over considerable distances. These snakes can trace prey even in the dark, and judge both the direction and the distance of the source of the heat. The pits may also aid in locating a recently killed prey, before its temperature has changed to that of its surroundings.

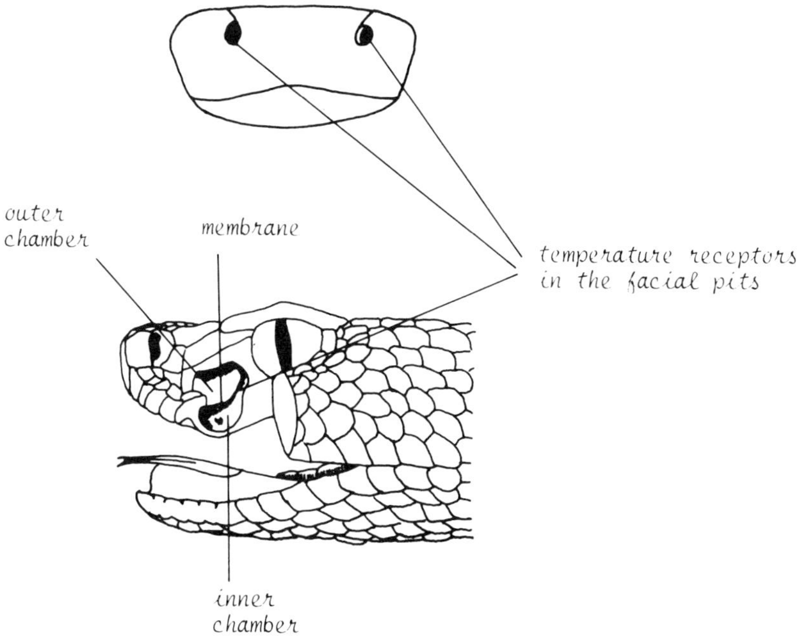

Figure 50. Heat sensitive receptors are highly developed in pit vipers.

The Venom Apparatus

In venomous snakes, the position and construction of the enlarged teeth (fangs) for transferring venom to the victim are, of course, very important. The fangs may be grooved or hollow and are always situated in the upper jaw. The fangs are connected, but not directly, to the venom glands, which are modified salivary glands of varying complexity in different genera.

The most primitive form of venom fang is found in the Opisthoglypha or back-fanged snakes, which embrace the colubrids (Figure 51). Their fangs, one to three, are grooved and placed at the rear of the upper jaw. The venom trickles into the wound made by

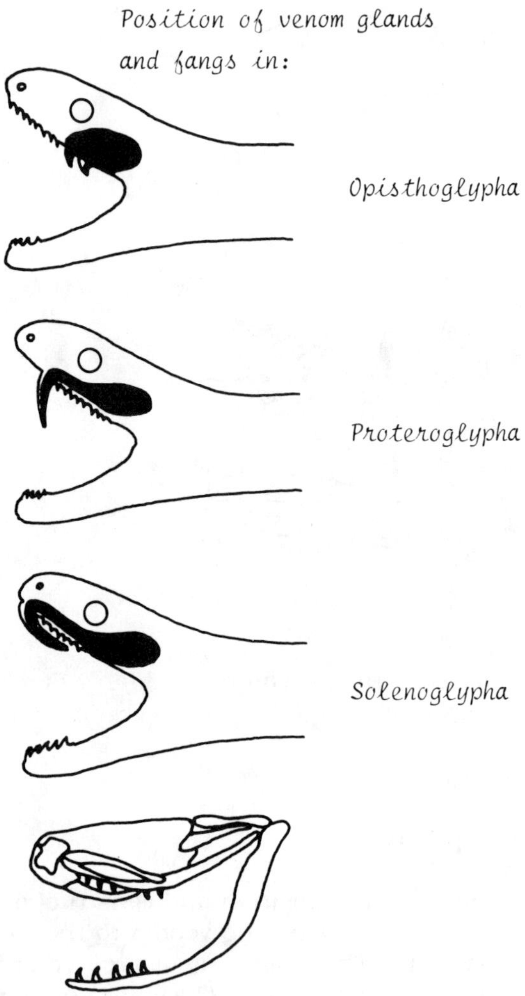

Figure 51. The venom apparatus of snakes.

the teeth. These snakes are seldom dangerous to larger mammals, including man, since the construction of the venom apparatus does not make it possible to inject venom in larger prey, and the venom itself is generally rather weak. However, as we shall see later there are a few exceptions.

A more efficient venom apparatus characterizes the Proteroglypha, which consists of elapids and sea snakes. Their fangs are grooved, nonmovable, and situated at the front of the upper jaw. These snakes also seize their prey and chew on it to introduce the venom. The frontal position of the fangs makes the apparatus much more efficient; many of these snakes have very potent venom and are dangerous to man.

The construction of the venom system has been optimized in a third group, the Solenoglypha, which consists of the true vipers and pit vipers. The fangs are situated at the front of the upper jaw but are now movable and can be erected when the mouth is open. The fact that the fangs are erected when the snake strikes, and can be folded back along the roof when not in use, makes it possible for these structures to be very long, about 25 mm for a large puff adder. If the long fangs are broken, replacement fangs are ready to take over the job. The fangs are canalized and the duct of a large venom gland expands into a small cavity in the gum, overlying the opening of the venom canal. Snakes of this group stab the prey with the erected fangs, while at the same time injecting the venom, and withdraw the fangs again immediately. The whole operation only takes a few seconds or less and the prey may travel some distance before it succumbs. However, the special sense organs of the viper make it possible to easily track down and find the victim.

The General Composition of Snake Venoms

All snake venoms are very complex and consist of numerous components. They contain several different proteins, many of which are important for digestion of food. Different peptides and polypeptides are responsible for the toxic properties. In addition, there are inorganic ions like calcium, magnesium, cobolt, nickel, zinc, and manganese. Some of the ions play a role in the activation of enzymes. For instance, zinc is essential for the function of certain proteolytic enzymes, some of which have hemorrhagic effects, causing capillaries to break open and blood to leak out in the tissues of the prey.

The venoms of snakes also contain several lower molecular weight compounds, e.g., nucleosides and biogenic amines. Acetylcholine exists in such a high concentration in venoms of African mambas, that it has been considered that it may affect acetylcholine-sensitive mechanisms of the prey. However, the functional significance of the lower molecular components in snake venoms is largely unknown.

A total of about 25 different enzymes have been identified in snake venoms (Table 4), but the number varies in different species. Proteolytic enzymes, which catalyze the breakdown of tissue protein, are abundantly present in viper venoms and in particular, in those of the pit vipers. Such enzymes are less common and sometimes absent in elapid and sea snake venoms. The amount of proteolytic enzymes often correlates with the ability of the venom to produce various kinds of tissue distruction in the prey. The molecular weights of the proteolytic enzymes range from about 20,000 up to 95,000.

In venoms rich in proteolytic enzymes, e.g., in rattlesnake venoms, there are components that affect the blood coagulation mechanisms of the prey. Among these components, the thrombin-like, and fibrinogenolytic enzymes, factor X activator, and platelet activating enzymes have attracted much interest in view of their potential clinical use and applicability as tools to elucidate the complicated mechanisms of blood coagulation. As pointed out earlier, hyaluronidase is an ingredient of most animal venoms, including the snake venoms. Hyaluronidase dissolves the intercellular matrix helping to speed up the spread of the venom. Collagenase is an enzyme which digests the collagen substance of connective tissues. This enzyme is primarily found in the venoms of true vipers and pit vipers, i.e., in venoms that

Table 4. Enzymes in snake venoms.

Proteolytic enzymes	Ribonuclease
Hyaluronidase	Deoxyribonuclease
Collagenase	a-amino acid oxidase
Phospholipases	Acidic phosphatase
Lactate dehydrogenase	Basic phosphatase
Phosphomonoesterase	5-nucleotidase
Phosphodiesterase	NAD-nucleotidase
Acetylcholinesterase	

catalyze the breakdown of tissue proteins. Acetylcholinesterase is a common constituent of elapid venoms and is also found in sea snake venoms, although it is absent in true viper and pit viper venoms. This enzyme catalyzes the hydrolysis of acetylcholine to choline and acetic acid, but its role in venoms is unclear. The possibility that the enzyme contributes to the neurotoxic effects of elapid and sea snake venoms has been discussed.

Besides proteolytic enzymes, snake venoms contain to a varying extent a number of well known enzymes such as ribonuclease, deoxyribonuclease, nucleotidases, amino acid oxidase, lactate dehydrogenase, acidic and basic phosphatases, etc. These and the proteolytic enzymes probably serve as digestive enzymes starting the process of digestion before the snake has eaten its prey. In comparison to the snake, the prey is often rather large and is swallowed in one piece. According to some investigations, the acitivity of digestive enzymes in the venom could facilitate penetration of the snake's gastric juice into the prey.

Among the different components in snake venoms a remarkable one is the nerve growth factor (NGF). It has been speculated that NGF could potentiate the toxic effects of the venom by stimulating cells of the prey, making them more vulnerable to the toxic components.

Among the toxic components there are two classes of neurotoxins, termed α-and β-neurotoxins. The so-called α-toxins are the better characterized of the two. They act postsynaptically by binding to the nicotinic acetylcholine receptors and by this prevent transmission of the nerve impulse. These postsynaptic neurotoxins are found in the venoms of elapids and sea snakes.

Certain polypeptides, which show similarities to phospholipase A2 enzymes, play an important role in the toxic effects of various snake venoms. Phospholipase A2-related components have been found in all snake families containing venomous species. Some of these polypeptides are responsible for the presynaptic effects of venoms from certain species of pit vipers, true vipers and a large number of elapids. These toxins have been termed β-toxins, and all appear to affect the release of acetylcholine from the nerve terminal. We shall later return to the chemistry and pharmacology of the postsynaptic (β-toxins) neurotoxins.

In the literature of snake venoms there is sometimes a trend to categorize the venoms into nerve venoms, heart venoms, necrotic venoms, etc. We know today that this classification is a too broad generalization, and that most snake venoms contain several toxic components which simultaneously affect different tissues. The toxic effect of a venom is to a large extent determined by the amount of the various toxic components which are accumulated in the different tissues. With some exceptions, for instance, the α-neurotoxins, snake toxins seldom show pronounced tissue specificity, but interfere with basic chemical mechanisms or cellular structures which are generally common for the different cell types. To illustrate this it can be mentioned that the β-neurotoxins of the Australian Tiger snake and of the Taipan not only cause neurotoxic symptoms but also have hemolytic and myonecrotic effects. The term β-neurotoxin as such may obviously be misleading when we discuss the venoms of these snakes. In the discussion that follows concerning neurotoxic, cardiotoxic, coagulation affecting, hemotoxic, hemorrhagic, and myonecrotic snake toxins, the reader should be aware that a specific toxin often has more effects than the name suggests.

Neurotoxins in Snake Venoms

Among the snake neurotoxins, the type which acts by binding to the nicotinic acetylcholine receptors is best known. This type comprises the so-called postsynaptic neurotoxins or α-toxins. Similar to curare, they prevent the acetylcholine-mediated impulse transmission. A great number of α-toxins from different elapids and sea snakes have been isolated and characterized. They are all basic and very temperature stable polypeptides. The amino acid sequence is known for several α-toxins, and in some cases also the three-dimensional structure (Figure 52). α-toxins of various species show differences but also pronounced similarities, with molecular weights ranging between 6,000 and 7,000 and constructed of 61 to 74 amino acids. The protein chain is stabilized by 4 or 5 disulfide bonds. These toxins have both a higher affinity to and specificity for the nicotinic acetylcholine receptor than curare. Just like curare they cause respiratory failure in animals, but at similar concentrations are considerably more toxic than curare. α-toxins have been used as pharmacological tools to localize, isolate, and investigate the properties of the nicotinic acetylcholine receptors. In this respect α-bungarotoxin from the Taiwanese krait (*Bungarus multicinctus*) has been of especial importance. It was isolated in the early sixties by Lee and collaborators. The toxin can be labeled with a radioactive isotope, e.g., I-125, and then be used as a sensitive marker for the receptors.

The so-called presynaptic neurotoxins, or β-toxins, are with respect to their chemistry and pharmacology much more difficult to survey and summarize than the α-toxins. These polypeptides are also basic and show temperature stability. They are at least twice as large as the α-toxins. The β-toxins are all related to phospholipase A2 and are composed of one or two peptide chains of 60–140 amino acids each. If there are two chains, they are held together by a disulfide bond, while the chains themselves are also stabilized by 6 to 8 disulfide bonds. The β-toxins vary more in structure and also in mechanism of action than the α-toxins do.

There is no quantitative correlation of the phospholipase activity of β-toxins with their toxicity. On the other hand, if the phospholipase activity is inhibited by blocking the essential amino acid histidine, the toxicity is also abolished. The enzyme activity and the toxicity, are both dependent on the presence of

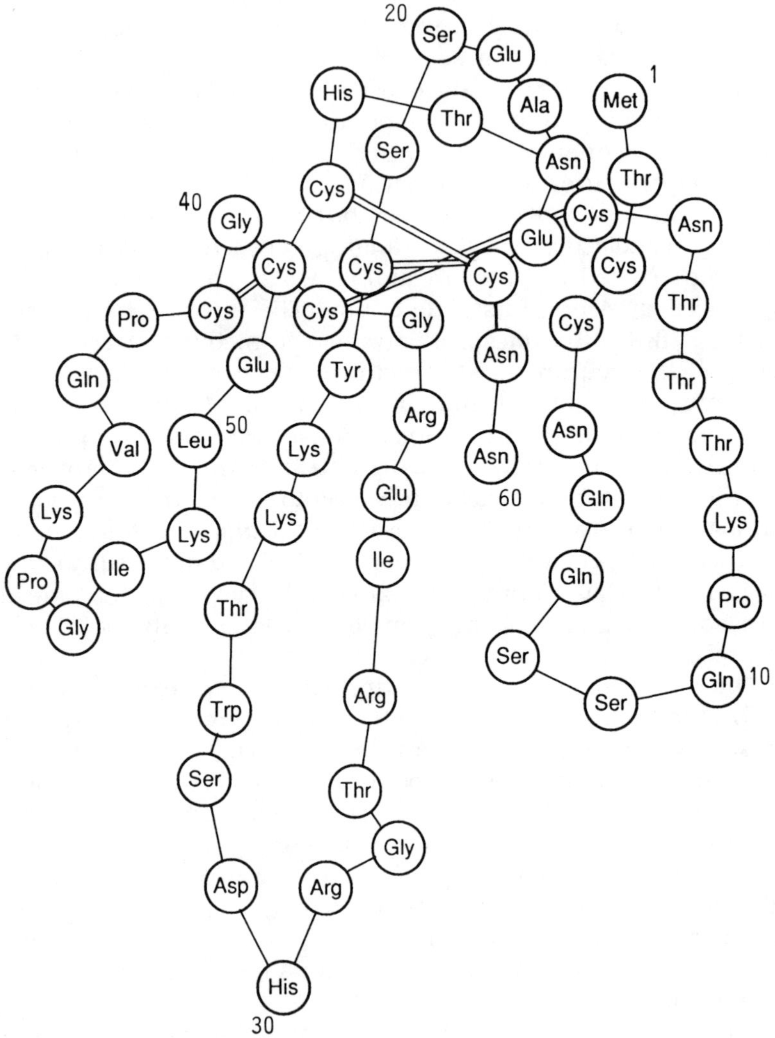

Figure 52. The chemical structure of the postsynaptic acting neurotoxin (α-toxin) of a sea snake, Lapemis.

calcium ions. The toxic effects of β-toxins appear to be enhanced by other components of the venom. This is best known for crotoxin the main toxic component of Brazilian rattlesnake (*Crotalus durissus terrificus*) venom. Crotoxin can be separated into one basic component, which carries phospholipase activity and is characterized by low toxicity, and into one associated acidic non-toxic protein. When the two components are allowed to act in combination, full neurotoxicity is achieved. A similar synergistic action of a basic phospholipase and an acidic inert component also appears to be typical of some other kinds of β-toxins.

The β-toxins all act presynaptically by affecting the release of acetylcholine via mechanisms which may be somewhat different for the different β-toxins. They mostly show very high toxicity and ultimately cause respiratory paralysis. Generally β-toxins cause the disappearance of the vesicles which contain acetylcholine, preventing the controlled release of acetylcholine and thus blocking impulse transmission. The action of β-toxins shows some resemblances to that of botulinus toxin. Four β-neurotoxins have been investigated especially thoroughly. Besides crotoxin from the Brazilian rattlesnake, they are β-bungarotoxin from the Taiwanese krait (*Bungarus multicinctus*), notexin from the Australian tiger snake (*Notechis scutatus*) and taipoxin from taipan (*Oxyuranus scutellatus*). We have previously become acquainted with presynaptic neurotoxins in other animal groups, such as the α-glycerotoxin from a polychaete (*Glycera convoluta*) and latrotoxin from the Black Widow spider (*Latrodectus mactans*). In contrast to the presynaptically acting snake toxins, these toxins inhibit acetylcholinergic transmission by triggering an initial release of the transmitter and preventing the uptake-storage mechanism of the vesicle from acting.

Besides the "classic" α-toxins other types of neurotoxins have been identified in other snake venoms. The mambas (*Dendroaspis*) belong to the Elapidae and are represented by four species, all living in Africa. Their venoms are highly complex and composed of about 30 different proteins. α-neurotoxins appear to be the only kind of toxin that mambas have in common with other snake neurotoxic venoms. The mamba venom contains toxins which

Table 5. Animal toxins which affect chemical synaptic transmission

Species	Toxin	Mechanism
Tedania ignis (Sponge)	Not characterized.	Inhibits evoked release of ACh. Reversible effect.
Tedania ignis	Not characterized.	Induces spontaneous release of ACh. Reversible effect.
Glycera convoluta (Polychaet)	α-glycerotoxin, globular protein. Mw = 300000.	Induces release of ACh. Reversible effect.
Glycera dibranchiata (Polychaet)	Not characterized.	Induces release of transmitters in different nerve terminals in PNS. Similarities to latrotoxin.
Cone snails	α-conotoxin. Mw = 1300.	Postsynaptic curare-like effect.
Cone snails	ω-conotoxin. Mw = 2600.	Inhibits Ca^{2+}-channels and prevents release of ACh.
Black Widow	Latrotoxin. Mw = 125000.	Causes massive transmitter release in nerve terminals in PNS. Ultrastructural changes.
Funnel-web spider	Atraxin. Mw = 15000–20000.	Causes massive transmitter release in nerve terminals in PNS. No ultrastructural changes.
Elapids, sea snakes and certain other snake species	Mw = 60000–70000.	Postsynaptic curare-like irreversible effects, so called α-toxins.
Elapids, pit vipers and a few vipers	Mw = 20500 (heterodimer), phospholipase activity.	Presynaptic so called β-toxins. Irreversibly affect release of ACh. Botulinum toxin-like effect.
Mambas	Dendrotoxin, toxin I. Mw=7000.	Inhibits K^+-channels and induces release of ACH.

have been shown to facilitate acetylcholine release and thus synaptic transmission at the neuromuscular junction, leading to excessive muscular activity, trembling, and fasciculations of the prey. The action can be ascribed to the selective action of special toxins (dendrotoxin, toxin I) on the voltage-dependent potassium channels. The toxins have molecular weights of about 7,000. These and some other special animal toxins have proven invaluable in characterizing the many different types of potassium channels and their roles in cell function.

In the animal kingdom a large variety of toxins have evolved which affect synaptic transmission, especially the acetylcholinergic type, which has a central role for several life-sustaining functions. In Table 5, a list of such toxins is presented.

Cardiotoxins in Snake venoms

A toxin isolated from Indian cobra venom in the late forties was called cardiotoxin since it could early be shown that it caused cardiac arrest when injected in experimental animals. Similar toxins have subsequently been isolated from the venoms of various cobra species. These toxins are also known as cytotoxins, direct lytic factors and membrane toxins. They are basic polypeptides which consist of 60 amino acids and contain 4 disulfide bonds. They show striking similarities to the α-neurotoxins but in spite of this do not have any pronounced affinity for acetylcholinergic receptors. The principal effects of cardiotoxins are those on excitable cells. They cause depolarization and contracture of cardiac, skeletal and smooth muscles, and depolarization and loss of excitability of nerves. At higher concentrations, they cause hemolysis of erythrocytes, which are not likely to be prime targets. The mechanisms of action are so far unknown. One suggestion is that cardiotoxins increase the permeability of cell membranes to inorganic ions by displacement of calcium from the cell membrane. Another possibility is that the toxins form ion-channels in the membrane, like melittin from the honeybee or jellyfish toxins.

Hemorrhagic and Myonecrotic Toxins in Snake Venoms

Bleeding, also called hemorrhage, and local tissue damage, are the most striking local symptoms after the bites of many pit vipers (Crotalidae) and vipers (Viperidae). In severe cases, hemorrhage may also occur in many internal organs. These toxic effects are caused by specific hemorrhagic and myonecrotic toxins, which are best known from studies of rattlesnake venoms. Six hemorrhagic toxins have been isolated from the Western diamond rattlesnake (*Crotalus atrox*) venom, and all show proteolytic activity. Each molecule contains one zink ion and the molecular weights vary from 24,000 to 68,000. These toxins break up the capillaries allowing non-coagulable blood to leak out into the surrounding tissues. It appears that hemorrhagic toxins contribute to the lethal action of rattlesnake poisoning to a much higher degree than the neurotoxins do. Hemorrhagic factors are present in the venom of the King cobra (*Ophiophagus hannah*) but have otherwise only been found in those venoms from pit vipers and true vipers.

Toxins causing tissue damage, or necrosis, are found to a varying degree in most snake venoms. Besides proteolytic and hemorrhagic components, the damage may often be caused by phospholipase A activity. Furthermore, there exist some types of toxins which specifically destroy muscles. These are called myonecrotic toxins, and are found in venoms from rattlesnakes and other pit vipers. One of the best known is called myotoxin *a*. It has been isolated from the venom of the Prairie rattlesnake (*Crotalus viridis*) (Figure 53). It is a basic protein having a molecular weight of 4,600 and lacking enzymatic activity. Myotoxin *a* binds specifically to the sarcoplasmic reticulum of the muscles. This leads to changes in ion permeability of the sarcoplasmic reticulum, known to be the

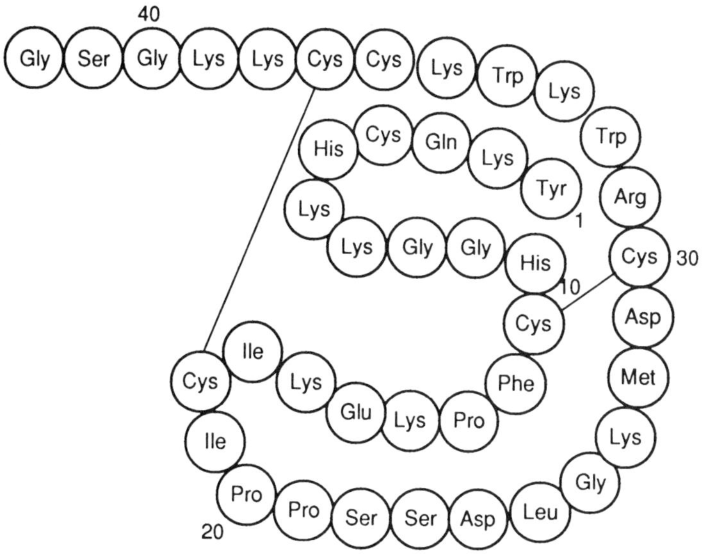

Figure 53. Myotoxin a.

important calcium regulatory system of muscles. This in turn leads to swelling, followed by disintegration of both the sarcoplasmic reticulum and the muscle fibrils.

Factors in Snake Venoms Which Affect Blood Coagulation

With the exception of venoms in sea snakes and some elapids, snake venoms contain various components that interact with the blood coagulation system. These factors provide unique tools for elucidating the very complicated cascade of events which underlie both the coagulation of blood and the dissolution of the blood clot. There are also good reasons to believe that such factors may become useful medically for the treatment of disorders in coagulation mechanisms. Such expectations to some degree have already been fulfilled.

These venoms act by preventing blood coagulation, which promotes bleeding of the prey. However, this process may initially be preceded by the effects of certain venoms which stimulate coagulation. In snake venoms, there may simultaneously exist factors which both stimulate and inhibit coagulation.

Most elapids contain anticoagulant factors. In cobras, this may be due to the presence of factors which destroy tissue thromboplastin, which has a role in the functioning of the clotting mechanism. In venoms of the common European adder (*Vipera berus*) and of some other viperids, there are components which inhibit the production or activity of the thromboplastin system.

Pit viper venoms are generally recognized to have significant effects on the blood. There are primarily three different activities that account for these effects on blood coagulation; thrombin-like enzymes, platelet aggregating factors, and fibrinogen and fibrin dissolving enzymes. Thrombin-like enzymes convert blood fibrinogen to fibrin, but in contrast to what normally takes place, very small intravascular clots, microembuli, are formed. These clots can next be dissolved by the fibrin dissolving enzymes. Since the fibrinogen has been converted to fibrin, which is now dissolved, further coagulation is prevented. The fibrinogen store of the prey is also destroyed by fibrinogenolytic enzymes, which further impair the blood clotting ability. The thrombin-like and platelet activating enzymes belong to the serine proteinase class. The fibrin and fibrinogen dissolving enzymes are metalloproteinases. These enzymes are of great interest both as research tools and as possible therapeutic agents in medicine. A thrombin-like enzyme from the Malayan pit viper (*Agkistrodon rhodostoma*) has already been used for the prophylactic treatment of patients disposed to blood clots. It is called Ancrod, and in Europe is commercially available under the name Arvin. This enzyme acts as an indirect anticoagulant by slowly converting fibrinogen into very small clots of blood, microembuli, which can next be eliminated from the circulation.

Besides the thrombin-like enzymes, there are other factors in pit viper venoms which increase the initial events in the clotting pathway by stimulating the aggregation of blood platelets. Such factors are crotalocytin (*Crotalus horridus*), convulxin (*Crotalus durissus*), aggregoserpentin (*Trimeresurus mucrosquamatus*) and thrombocytin (*Bothrops atrox*). These stimulators are polypeptides or glycoproteins, with molecular weights which vary from 35,000 to 80,000.

Other Toxic Factors in Snake Venoms

Snake venoms are as complicated as their physiological effects. Besides the previously mentioned neurotoxic-, cardiotoxic-, hemorrhagic-, myonecrotic-, and blood coagulation- effects, snake venoms primarily or secondarily affect several other functions of the prey. We shall only mention a few additional toxic effects, and the reader is directed to apply to more thorough-going reviews of this interesting field.

The venom of most venomous snakes, possibly with the exception of sea snakes, produces an immediate lowering of the blood pressure and a physiological shock, which for pit vipers, and true vipers in particular, are the most important reasons for the lethal effects of their venoms. With respect to pit vipers, the blood pressure fall appears to be explained by the dilatation of the blood vessels of the skeleton muscles, and for true vipers mainly to be due to the dilatation of the vessels of the internal organs. The fall in blood pressure or shock is caused by a decrease in the circulating blood volume due to an increase in capillary permeability. Some venom components appear to produce decrease in blood pressure by a direct effect on the peripheral blood vessels. These factors are still little known. From rattlesnake venoms, a basic polypeptide (mol. weight = 4,490) has been isolated which causes a large fall in blood when injected in experimental animals.

The effects on blood circulation might also be due to pharmacologically active substances which are released from the tissues of the prey by the action of venom components such as phospholipases and proteinases. The released factors could be bradykinin, histamine, serotonin, prostaglandins and others which decrease blood pressure and increase the permeability of the blood vessels. The pharmacological effects are not limited to the blood circulation but may also be supposed to affect endocrine glands, metabolic events, etc.

The hemolytic effects of snake venoms have attracted much interest. However, hemolysis, that is, the destruction of red cell membranes by certain venom components, is not considered to play a dominant role in the lethal action of the venom. Direct hemolytic activity is mainly found in the venoms of elapids and to somewhat lesser extent, in the venoms of pit vipers and true vipers. The hemolysis may be due to a direct action of phospholipase A on the red cell membrane. It may also be due to the catalysis of lecithin in the cell membrane by phospholipase A, with the subsequent formation of lysolecithin. Besides phospholipase A, a direct lytic factor (DLF) has been found in cobra venoms. DLF is a protein identical to cardiotoxin (see earlier text). The hemolytic effect requires a much higher DLF concentration than the cardiotoxic DLF concentration does. However, DLF in small amounts can increase the hemolytic effect of phospholipase A. Here we again have an example of a synergistic effect. These are common in animal venoms, for which it is often true that the whole is much larger than the sum of the parts.

How Dangerous Are Snake Venoms?

Annually, many humans are bitten by venomous snakes. However, most bites are "startled" bites or "defensive" bites and the snake often does not inject venom even though several lethal doses of venom may be present in the venom glands at the time of the bite. Only about a half of bitten humans may experience poisoning symptoms and of these, only a small percentage have a fatal outcome. Of course children and elderly people run the greatest risk of being severely poisoned.

There are no worldwide statistics available, so in order to form an opinion about the health hazards of snake bites the results of separate investigations from different regions of the world must be studied. In a study of 897 reported bites of true vipers and elapids in Natal, Indonesia, it could be established that only 11% were severe and 2.4% were followed by death. Each year there are about 6,000 to 7,000 snake bites in North America, mainly caused by rattlesnakes. Less than 1% of the cases which receive medical treatment have a fatal outcome. In 1979, there were 5 deaths from venomous reptile bites in the United States. It should be pointed out that many snake bites are never reported, because the bitten persons do not consider that medical attention is required. Worldwide, deaths have been estimated from 30,000 and up to 40,000 in some older reports. There is no reliable information and even the figure of 30,000 could be an overestimate. The greatest number of fatal snake bites is reported from Southeast Asia, with the heaviest mortality in Ceylon, India, Pakistan, and Burma. However, there is no doubt that the incidence of snake bites in developing countries decreases, since the snake population also usually diminishes.

The improved availability of antiserum has contributed to the reduction of the health hazard during recent years. Antiserum is generally produced by immunized horses. Because there are biochemical differences in the venoms from the various species, a polyspecific antiserum, which can be used against several different venomous snakes, is preferred for use in snake-rich regions of the world. A good reason for using polyspecific antiserum is also that in most cases of snake bite, the snake is neither seen nor identified.

Some years ago a team of missionary doctors from Ecuador described a new method to treat snake bites. In their study a series of high-voltage, direct-current shocks applied directly to a bite site prevented mortality and pain among members of a population of Ecuadoran Indians where death from snake bite was common. The doctors suggested that the electric current may work by interfering with the electrostatic forces that hold the venom's proteins in their functional shape. In addition, the electric shock might cause muscles around the bite to spasm, cutting off circulation and isolating the poison. Other studies suggest that electric shock may also be effective for victims of scorpions, stingrays, bees, and wasps. However, laboratory experiments using electric shocks under controlled conditions have so far been ineffective in treating the lethal effects of snake envenomation in mice and rats.

Several plants are also claimed by popular belief to counteract the effects of snake bites, and recent research into this possibility has been very positive. Extracts of a herbaceous plant (*Eclipta prostrata*), known in Brazil as erva-botao, have been shown to inhibit the lethal and myotoxic activities of South American rattlesnake venom. The active "anti-snake" constituents of the extract are now being identified.

Several factors affect the toxicity of the venom, such as the age, size and condition of the snake. There may also be seasonal, and even geographical differences, in the toxicity of the same species. The amount of venom increases the longer the time that has elapsed since the preceding bite was given. Various animals differ markedly in their susceptibility to snake venoms. For example, cats and frogs are much more resistant to cobra venoms than most rodents. The most familiar example is perhaps the Asian mongoose. Many of us have read of the fights of the brave mongoose Rikki Tikki Tavi against the cobra. It was once believed that the mongoose returned to the jungle after being bitten to eat a plant that contained an antidote to the venom. The fact is that the mongoose has a natural resistance to cobra venom. Horses have also acquired some tolerance towards various snake venoms. The European hedgehog (*Erinaceus europaeus*) is known to kill and eat vipers. The hedgehog has been found to tolerate 40 times as much venom of the European viper (*Vipera berus*) as a guinea pig of the same size. It has been suggested that the hedgehog plasma contains certain proteinase inhibitors, which inhibit the venom's hemorrhagic metalloproteinase. Antihemorrhagic proteins have also been isolated in serum of the wood rat (*Neotoma micropus*), which can tolerate 140 times as much venom from the Texas rattlesnake (*Crotalus atrox*) as a laboratory mouse. It is not surprising that antihemorrhagic factors also have been observed in the sera of various snakes. The mechanisms of venom resistance is an interesting but so far little exploited field. However, the venom neutralizing activity of the sera of certain animals has sparked an interest in their possible use in treating snakebites.

Colubrids (Colubridae)

In this and some following sections we shall briefly give examples of venomous snakes from the different families, without becoming absorbed in the toxinological aspects, which have been discussed earlier.

About 80% of all known snakes belong to the family Colubridae. From the toxinological point of view, the family can be divided into two groups, the Aglypha and the Opisthoglypha. The Aglypha lack fangs and are harmless snakes. The majority of colubrids are in this group. Even if some aglyphous snakes may have toxic salivary gland secretions they have difficulty in introducing it into larger prey and are of no threat to humans.

In contrast, opisthoglyphous colubrids are venomous and have one or more pairs of grooved teeth in the back of the upper jaw. Here we find some snakes which are also deadly poisonous for humans. The most dangerous and famous snake is the boomslang (*Dispholidus typus*). It lives in Africa, mainly in the savannah and the bush. It is a tree-living, full meter-long snake with a greenish camouflage color. It moves faster among the branches of the trees and bushes than on the ground. The venom is remarkably toxic and has a similar mechanism of action as viper venoms. This snake feeds mainly on lizards, in particular on chameleons. It is not very aggressive and, moreover, its mechanism for injecting venom is rather inefficient. Still, it has been responsible for a number of fatalities among humans in situations when it has been frightened.

Another potentially dangerous colubrid for man is the bird snake (*Thelotornis kirtlandi*), which lives in Africa, in the southern Sahara. Like the boomslang, it spends most of its life in the trees, its color providing a perfect camouflage. The average length of the bird snake is 120 cm. If it is alarmed, it takes up a threatening position by blowing up the throat and the forepart of the body. The venom shows similarities to that of the boomslang.

Elapids (Elapidae)

The elapids (Elapidae) include the most poisonous and dangerous snakes in the world, and are represented by about 170 species. Elapids occur on all continental tropical and subtropical landmasses, and they are the dominant venomous snake in Australia. They also occur throughout the East Indies. They are proteroglyphous snakes, characterized by having nonmovable, rather short fangs situated anteriorly in the upper jaw and connected to large venom glands. Some of the most famous and feared snakes are elapids, for instance, the cobras, the mambas and some of the Australian snakes.

The Indian or Spectacled cobra (*Naja naja*) is perhaps the most well known of all snakes. It is the only Asian cobra belonging to the genus *Naja*. Among other things, the characteristic spectacle-like marking on the neck has contributed to its fame. When it is threatened, the neck expands to the shape of a shield and the spectacle is clearly discernible. The poor creatures which are used by snake-charmers and in exotic dances are prevented from biting, either by sewing the snake's mouth shut or by a vasoectomy to eliminate the channeling of the venom of the gland to the fang. The Indian cobra may be up to 1.5 meters in length and mainly feeds on rats, mice, frogs, and birds. It goes hunting in the dusk, when humans are generally bitten. Cobra bites are common in India and about 10% of the bites have a fatal outcome. The dominant toxic effect is due to the postsynaptically acting neurotoxins (α-toxins).

The King cobra (*Ophiophagus hannah*) is the largest of the venomous snakes, attaining a length of 5 meters. It occurs rarely in India and in large parts of southeast Asia. It inhabits sparsely settled regions and therefore seldom encounters human beings. It feeds almost exclusively on snakes, harmless as well as venomous ones, is active during daytime and retires if humans arrive. Rumors of its aggressiveness and tendency to unprovoked attacks are exaggerations and myths. However, because of its enormous size and strong venom it is potentially one of the most dangerous snakes in the world, capable of killing a man in just a few minutes.

The Egyptian cobra (*Naja haje*) is found in territories extending from northern Africa, including all the countries bounding the Sahara, southwards throughout the eastern half of the continent to Natal and Transvaal. Large specimens are up to 2.5 meters in length. They feed on lizards, birds, toads, and frogs, and birds appear to be very sensitive to their venom. They do most of their hunting for food at night and are great raiders of poultry runs. In spite of the impressive size and rich venom secretion, they appear to cause few human deaths. A famous victim of this snake's venom may have been Cleopatra, who, according to the legend, committed suicide by means of a cobra bite.

The Black-necked spitting cobra (*Naja nigricollis*) and South African spitting cobra or Ringhals (*Hemachatus haemachatus*) are common in large areas of Africa. Their venoms are dominated by potent neurotoxins. These species have adopted the strange habit of spitting the venom at the attacker, aiming at the eyes with a high accuracy. When the venom is to be ejected, the surrounding muscles squeeze the glands, forcing the venom into the duct and down through the fang and out of the orifice. The venom is discharged as two fine jets at an enemy up to a distance of 1.5 to 2.5 meters. The venom penetrates into the eyes and causes serious lesions.

The mambas (*Dendroaspis*) comprise four African species existing throughout tropical Africa, from Somaliland on the east to Senegal on the west, and southwards into southern Africa. Three of the species are the so-called green mambas; Green mamba (*D. augusticeps*), Jamesons mamba (*D. jamesoni*), and Guinea mamba (*D. viridis*). They hunt in the trees and have an effective camouflage color, which makes them almost impossible to discover in the foliage. Much more feared is the notorius Black mamba (*D. polylepis*), the largest venomous snake in Africa with an average length of adult specimens of about 3.5 meters. In contrast to the green mambas, this snake prefers at ground living, obtaining protection in mounds of stones and cavities. Its diet is confined to birds and small mammals, and has an exaggerated reputation of being extremely irritable and aggressive. This, in combination with its impressive size and rapid speed of movement, places the Black mamba among the leaders of feared snakes. Being the "King" of African snakes, comparable only to the Australian Taipan, the King cobra in Asia and perhaps the Fer-de-lance in the Americas, it has been the subject of many legends and myths.

Coral snakes are beautiful, being brilliantly colored with bands of yellow, red, and black that may function to advertise the venomous nature of the snakes. About 40 species are known, all of which live in the warmer areas of the American continents. Among them, there are both highly venomous species (*Micrurus*) and some which are completely harmless. The latter gets protection by mimicking the warning coloration of the venomous ones. The coral snake venom contains very potent neurotoxins. The fangs are small and the snakes are not very aggressive. Still, there are several reported cases where humans have died from bites.

There is a very high proportion of venomous and dangerous snakes in Australia, in comparison with other countries. They are all elapids and some members are the most feared of all the snakes. Among the latter, is the strongly built Taipan (*Oxyuranus scutellatus*) with an average length of 2.5 meters. It has very long fangs and produces large amounts of an extremely potent venom containing a presynapically acting neurotoxin (β-toxin) and various necrotic toxins. It caused many human deaths before an antivenom became available in 1955. Fortunately, the Taipan is found only in the most northern parts of Australia, where the population density is low. Since it is also a rather shy animal, the number of accidents is small. It hunts during the day and feeds mainly on small rodents.

The Tiger snake (*Notechis scutatus*) is another snake of bad repute. It has a tiger-like color with yellowish bands against a pale-brown to sometimes black background. It is solidly built with a full grown length of 1.3 meters. Its venom is similar to that of the Taipan's and is present in large quantities. Fortunately, its fangs are rather short and therefore the injected amount of venom is relatively small. Moreover, these snakes are not aggressive unless provoked. However, they live in southeastern Australia, near areas of dense population, and this snake is the most common cause of serious snake bites in Australia. Persons bitten need immediate treatment and antivenom.

Among other notorious Australian venomous snakes one can mention the Death Adder (*Acanthophis antarcticus*), which is distributed in bushland throughout most of Australia. It looks like a viper with a broad head, a thick neck and a fat body, and is about one meter or shorter in length. The short tail ends in a small sting-like tip and many incorrectly believe that this tip is used to deliver a venomous strike. During hunting, it often twitches the tail, a behavior which can be seen in several other snake species and which may be a way of luring the prey. Its venom is extremely toxic but it does not bite humans unless provoked.

A group of snakes called kraits live in southern Asia. Their venoms contain highly potent neurotoxins. In the venom of the Taiwanese krait or the Many banded krait (*Bungarus multicinctus*) both presynaptically and postsynaptically acting neurotoxins have been identified (β- and α-toxins). The Common krait (*B. caeruleus*) is found in India and Ceylon and is about 1 meter in length. Its venom is much more potent than that of the Spectacled cobra. Fortunately the Common krait retreats when it senses human approach and accidents are therefore not so frequent.

Sea Snakes (Hydrophiidae)

The sea snakes (Figure 54) are probably closely related to the elapids. Since they are highly specialized for an aquatic life they are generally arranged in a separate family. It consists of about 50 species, which are widely distributed in the South Pacific and the Indian Ocean, being particularly abundant around Australia and in the southeastern Indian Ocean. One species (*Pelamis platurus*), which has been found hundreds of miles from land, occurs from

Figure 54. A sea snake.

southeastern Africa and Madagascar across the Indian and Pacific oceans to the western coast of tropical America. Another species (*Hydrophis semperi*) has adapted to fresh water. It lives in a lake on the island of Luzon in the Philippines. All other species are inhabitants of marine waters. The average length of most species is about 1 meter or somewhat less. They are excellent swimmers and have a flattened, paddle-like tail which forces them forwards. They have valvular nostrils to prevent the entrance of water. The lung is very long and extends to the base of the tail. The marine sea snakes have, like certain sea birds, salt glands for the excretion of excess sodium chloride. In the sea snakes, the salt glands open into the mouth cavity, from which the salt is expelled.

The sea snake venoms contain very potent postynaptically acting neurotoxins (α-toxins). Some species have also been shown to contain a presynaptically acting neurotoxin (β-toxin). On a weight basis, the sea snake venoms appear to be more toxic than the cobra venoms, but they are also produced in smaller amounts. Like the elapids they are proteroglyphous snakes. Sea snakes feed mainly on fish and particularly on eels. They avoid swimming people and accidents mostly occur when handling fishing nets.

Thousands of sea snakes are caught annually and shipped to various countries, especially to Japan, where the meat is eaten fried or smoked.

Vipers (Viperidae)

The true vipers are represented by about 40 species distributed in Europe, Africa and large parts of Asia, but are absent on the American continents and in Australia. The true vipers, like the pit vipers, are solenoglyphous snakes, characterized by having long, movable fangs on the upper jaw that rotate back up into the mouth when not in use. Since the fangs protrude out of the mouth, solenoglyphs may strike an object of any size. This is the most developed and efficient construction for venom injection (see earlier text). After a strike, often delivered in a fraction of a second, the snake may have to leave the escaping prey. Even if snakes in this group tend to be sluggish and heavy bodied, they have no difficulties in tracking down and finding the victim when the venom has exerted its actions. It then carefully investigates the corpse with its sensitive, forked tonque, before proceeding to swallow it from the head end. The venoms of vipers in general have necrotic and hemorrhagic effects, and more specifically, affect the heart and the blood circulation. A few species also have neurotoxins, for example, Russell's viper (*Vipera russelli*) and the sand viper (*Vipera ammodytes*). The vipers live on the ground and most species avoid water.

The most widely distributed and commonest venomous snake of Europe is the Adder (*Vipera berus*) (Figure 55). This animal inhabits a vast area stretching from western Europe to eastern Asia as far as China, and in Scandinavia up to the Arctic circle. It prefers dry surroundings but may, in contrast to most other vipers, be seen near lakes and swims with ease. Its maximum length is about 1 meter. Its bite is considerably less dangerous to humans than bites of several other vipers. A slightly more toxic relative (*Vipera aspis*) is found in southern and southeastern Europe.

Figure 55. Adder, Vipera berus.

The sand viper (*Vipera ammodytes*) becomes somewhat longer than the common Adder (*Vipera berus*). It ranges from the Balkans through Turkey to the Caucasus. It has a very efficient venom apparatus, the most potent venom among European vipers, and its bite may be dangerous to humans. Its venom is used for the production of antivenins against other European vipers.

Russell's viper (*Vipera russelli*) is found in the Indian subcontinent, in southeast China extending to Taiwan and in parts of Indonesia. It is solidly built and measures 1 meter or somewhat more in length. It can inject large quantities of a very potent venom and is a feared snake. Russell's viper is the most frequent cause of fatal bites in Thailand and also a large problem in India.

The Puff-adder (*Bitis arietans*) is probably the most common and widespread snake of the African continent. The average length is about 1 meter, but specimens of 1.5 meters or more have been found. When disturbed, it inflates itself with air, letting it out again in loud hisses or puffs as a warning that it is about to strike. It is a solidly built snake; the fangs are relatively long and able to inject large amounts of a very potent venom. The Puff-adder is normally a sluggish snake, relying on its cryptic coloring to escape notice. For this reason and since it is so common in certain areas, it is one of the most dangerous snakes in Africa both to humans and to grazing stock. This snake is probably responsible for more accidents than all the other venomous African snakes. Among other African vipers, one can mention the Horned adder (*Bitis caudalis*) and the Gaboon viper (*Bitis gabonica*). The latter is a heavy and stout snake, measuring about 1.5 meters in length. It has unusually long fangs and a very potent venom. Fortunately it is said to be a good-natured creature and is unlikely to bite unless trodden on or otherwise directly molested.

Pit Vipers (Crotalidae)

The pit vipers (Crotalidae) consist of about 140 species, most of which are found on the American continents and the others are widely distributed in large parts of East Asia. They differ from their closest relatives, the true vipers, by the presence of heat-sensitive facial pits on each side of the face between the eye and the nostril. The pit organs are highly sensitive to infrared wavelengths between 15,000 and 40,000 Ångström and appear to be able to detect temperature differences of 0.2 degrees centigrade or less. The pit organs also provide a stereoscopic "vision" of the heat source and interpretation of the distance to the source. When a rattlesnake strikes, the direction of the strike appears to be guided by the infrared radiation of its prey, which usually is a small warm-blooded animal. Like true vipers the pit vipers are solidly built and are solenoglyphous snakes with an advanced apparatus for venom injection. The venom produces a variety of effects including necrotic, hemorrhagic, and blood pressure effects. A few species, the Cascavell (*Crotalus durissus*) and the Mojave rattlesnake (*Crotalus scutulatus*), also have presynaptically acting neurotoxins (β-toxins).

Some well-known pit vipers belong to the genus *Agkistrodon*. One species, the Halys snake (*A. halys*), is found in East Asia and partially in Europe. It is a rather harmless snake and normally not deadly for humans.

The water moccasin (*A. piscivorous*) occurs in the southern United States. The buccal cavity is white so this snake is also called the "Cottonmouth." It normally measures about 1 meter in length and is a semiaquatic species living near lakes and water pools, feeding on fish and frogs. Bites may cause human deaths and often produce prolonged local tissue injuries. However, with the use of different sounds and movements it usually forewarns that it feels threatened.

The Malayan pit viper (*A. rhodostoma*) has been mentioned earlier as an example of venomous animal which has had medical importance. In this case, it is a thrombin-like protein in the venom which has been used as an indirect blood anticoagulant for treatment of certain patients. This snake is not quite 1 meter in length and is found in Indo-China and in the Malay Peninsula and surrounding archipelago. It gets on well in densely populated agricultural districts and since it also is a nervous creature, several humans are bitten each year. There are rather few deaths due to this snake but slow-healing tissue injuries are common. The latter are usual after pit viper bites.

The Barba amarilla or Fer-de-lance (*Bothrops atrox*) is widely distributed in northern South America and often occurs near human settlements. Its robust body may reach 2 meters in length. It has a nervous behavior and can be very aggressive when disturbed, injecting large quantities of a very strong and quick-acting venom. It is, and rightly so, a very feared snake and belongs to the same "superclass" as the King cobra, the Black mamba and the Taipan. It is supposed to kill more people in South America than any other snake in this part of the world.

The Bushmaster (*Lachesis muta*) is, together with the King cobra, among the largest venomous snakes. Large specimens can measure 3.5 meters in length. They live in the forested country of southern Central America and the tropical regions of South America. The Bushmaster is a quiet and shy snake and not very common. It injects large amounts of a rather weak venom. When disturbed by humans it causes a noise that recalls the rattle of the rattlesnakes, the sound being produced on a suitable substratum with the scales on the underside of the tip of its tail. Its modified tail tip is thought to be an evolutionary precursor to the rattle of rattle snakes. There are few human deaths due to bites of this snake. During 1987 the Brazilian Ministry of Health attributed about 2% of the 19,000 reported snake bite accidents in Brazil to the Bushmaster.

The most well-known pit vipers are probably the rattlesnakes (genera *Crotalus* and *Sistrurus*). They are also regarded as the most developed or advanced of all snakes and are peculiar to the New World. There are about 30 rattlesnake species widely distributed in North, Central and South America. Most species occur in the southwestern United States and in northern Mexico. The rattles of rattlesnakes represent an effective warning device formed by modified scales. The terminal tail scale of newborn snakes is shed in the usual way with the rest of the skin during their first molt, but is not sloughed at the second and subsequent molts. In this way a long chain of scales or segments is produced, consisting of terminal scales shackled on to one another anteriorly and forming a rattle. The hindmost segments of the rattle are the oldest and are lost in the course of time. From 2 to 5 segments are added to the rattle annually. In spite of this, the rattles of most adult rattlesnakes contain only 5 to 9 and seldom more than 14 segments. When a rattlesnake is disturbed, it indicates irritation by rapid lateral vibrations of the tail, thus creating a very characteristic warning signal. The rattling seems to prevent small carnivores from attacking the snake.

The Prairie rattlesnake (*Crotalus viridis*) is found in the western United States and in southern Canada. The mature snakes vary in length, from 60 cm up to a maximum of 130 cm. They are common in the prairies, chaparral areas and in grassland, but are also found in the mountains into the pine zone. They hunt both

Figure 56. Cascavell, Crotalus durissus, is one of the most dangerous snakes.

during night and daytime and are therefore often seen. The danger of the prairie rattlesnake seems to be popularly overestimated. Even if the venom is quite toxic, humans are seldom fatally bitten.

The Timber rattlesnake or Banded rattlesnake (*Crotalus horridus*) measures about 1 meter in length. It lives in the wooded country of central and eastern United States. It is a common species even if it is constantly being pushed away from its natural habitats by man's activities. It is a moderately dangerous species.

Considerably more dangerous is the Texas rattlesnake or the Western diamond rattlesnake (*Crotalus atrox*). The latin name "atrox" means cruel. It is found in dry lowland terrain from the southern United States throughout Mexico. It generally measures from 75 cm to 180 cm in length. Like most rattlesnakes, it is probably most active at dusk and at night, but often also hunts in the daytime. Being a rather aggressive, widespread and large snake, with large fangs, more people in North America have suffered from its bite than from that of any other snake species.

The Cascavell (*Crotalus durissus*) is distributed over the dry country from southern Mexico, through Central America, northern and eastern South America to south of the Amazon basin. It is readily provoked and has a highly toxic venom. It can be up to 1.5 meters in length and is considered to be one of the most fearsome of the venomous snakes (Figure 56). A bite requires immediate medical treatment.

21

Mammals (Mammalia)

The mammals (class Mammalia) probably evolved from a group of primitive reptiles (Synapsida), which can be traced back some 200 million years. The first true mammals began to appear during the Jurassic period about 150 million years ago. Mammals are warm-blooded or homeothermic animals, with body temperatures which may vary from 30 to 40 °C, and for the majority of species are about 37 °C. With the exception of the monotremes, which lay eggs like their reptilian ancestors, the young of mammals are born alive and are nourished by milk. Mammals are also highly percipient animals with large brains. About 4,000 species are known. With the exception of the insects, the mammals enjoy a wider distribution on the earth than any other animal group and contain species which have adapted to live in the water, in the air, in a cold arctic climate, and in tropical heat.

Only a few mammalian species have developed a venom apparatus. They are found among the more primitive mammals and include the platypus, the spiny anteaters, the solenodons, and certain shrew species.

Platypus (*Ornithorhynchus anatinus*)

The platypus and the spiny anteaters are the only representatives of the monotremes (Order Monotremata). Besides their egg-laying habits, they have many other features in which they resemble the reptiles. They are not strictly homeothermic but can vary their body temperature between 28 and 35 °C by influencing the surroundings. The platypus (*Ornithorhynchus anatinus*) is the only existing species in its family. Because of their peculiar appearance and unique evolutionary position, these animals have been subjected to many studies and there is a vast literature about them (Figure 57). The platypus is found in streams and lakes in eastern Australia and in Tasmania. A fully grown male is 50 to 60 cm in total length, about one-quarter of which is tail, and it weighs about 1.8 kg. The female grows to about two-thirds of the size of the male. The body is covered with dense hair. Especially conspicuous is the muzzle, formed by the jaws and resembling a duck's bill.

In spite of all the studies of this strange animal, there is still much to learn about the platypus and particularly about its venomous properties. The venom apparatus is found only in the male. It is a movable 15 mm long, horny and conical venom spur, found on the inside of each hind leg near the heel. The spur is enclosed at the base in a fleshy sheath and is normally enfolded into the skin, but is extended when required. Each spur is connected by a duct to a venom gland, the size of which varies with the season and attains its maximum size in the breeding season. The gland is whitish and kidney shaped and is actively secreting only during and near the breeding season. In view of this and because the venom apparatus is confined to the male, it seems likely that the main use of the apparatus would be in combat between males for territory or females. Some observations suggest that the venom could be used to immobilize larger prey.

There are few reports of attacks on man and none of fatal envenomation. A strike seems to cause an immediate and intense pain, which may last up to a day. There is also swelling at the wound. Symptoms of shock, such as weakness, may also occur. Some studies with experimental animals suggest that the venom

Figure 57. The platypus is a mammal with several reptile-like properties.

has hemorrhagic and cytolytic effects and also causes a fall in blood pressure. Neurotoxic effects have not been observed. The chemical composition of the venom and its mechanisms of action are not very well known. There seem to be certain similarities to the venoms of true vipers.

Spiny Anteaters (Tachyglossidae)

Spiny anteaters consist of two genera, each with one species and some subspecies. One is the swift tongue anteater (*Tachyglossus aculeatus*), which is rather common in most of Australia and Tasmania and parts of New Guinea. The other species is the tongue anteater (*Zaglossus bruijni*), which is confined to New Guinea and the surrounding islands. These animals differ from the platypus in general appearance, being flatter with long beak-like muzzles. At the end of the muzzle is a small mouth that opens to allow movement of the long extensile tongue for catching insects. *Tachyglossus* males are about 6 to 7 kg in body weight, while *Zaglossus* males are larger and may weigh 15 kg or more. Their venom apparatus and the toxic effects are similar to those of the platypus. Much work remains to be done on the venomous functions of these animals.

Figure 58. Cuba solenodon becomes about 50 cm in length and has a toxic saliva.

Solenodons (Solenodontidae)

There are two solenodon species, which form the smallest family (Solenodontidae) within the order Insectivora, a very primitive order of the mammals. The solenodons are squirrel-sized animals, approximately 50 cm in length, about half of which is tail. The muzzle is long and cylindrical. They have powerful claws, which have adapted to digging and tearing apart rotten logs. The two species are the Haitian solenodon (*Solenodon paradoxus*) and the Cuban solenodon (*Atopogale cubanus*), (Figure 58). They are only found in these islands, are rare and are also under threat of extermination.

The venom apparatus is best known from studies of the Haitian solenodon. Its submaxillary glands are large and have ducts that open near the base of the large, deeply channeled second incisor. When the animal bites its prey, the toxin-containing saliva of the glands is probably drawn upward through the teeth channels into the wound by capillary force. Since the existence of the species is threatened, there are few studies of the properties of the venom. Injections of the venom in experimental animals cause paralysis and convulsions. The venom probably contains neurotoxins.

Shrews (Soricidae)

This family (Soricidae) is represented by about 250 species, which are divided into 21 genera. The shrews are small mouse-like animals of various types, some terrestrial, others aquatic, found throughout the world. The fur is usually short and velvety. The nose is long and pointed and the eyes are very small. Like other small mammals the shrews have a very high metabolic activity and for survival are dependent on continuous access to food. During a 24-hour period several species are supposed to consume an amount of food which corresponds to their own weight or more. They feed mainly on insects, earthworms, frogs and other small animals. Some species are equipped with an efficient venom apparatus and may attack and kill rodents of their own size or even larger.

Two species are especially known to produce a venomous saliva. They are the European water shrew (*Neomys fodiens*) (Figure 59) and the American short-tailed shrew (*Blarina brevicauda*). The venom apparatus resembles that of the solenodons. It consists of two submaxillary glands from which large quantities of toxic saliva are secreted via ducts, which open near the base of the first incisors. In contrast to the solenodons, there are no teeth channels but concave inner surfaces which collect and serve to transfer the venom to the bite wound. Shrew bites may be rather painful but otherwise produce little effect in man. In experimental ro-

Figure 59. The water shrew belongs to the smallest (6–10 cm) venomous mammals.

dents and in natural prey, the venom is very effective, causing neurotoxic symptoms with convulsions and death occurring. The reactions are similar to those produced by solenodon bites. It is likely that the venom apparatus is of great biological significance to these species in obtaining food. Some data suggest that the major use of shrew venom is the immobilization of insect prey for consumption by the shrew at a later time.

The pharmacology of the venoms of solenodons and shrews, members of primitive insectivores, shows certain similarities to that of the elapids, whereas the construction of the venom apparatus resembles that of the beaded lizards (Heloderma). Whether or not these facts are of any relevance for phylogenetic reasoning is unknown.

22

Our Fear of Venomous Animals

Many persons feel an unreasonable fear of venomous animals, in particular of snakes, spiders and scorpions, the horror of snakes being the most common. In spite of the fact that the snake may be at a safe distance, panic is aroused, when suddenly the snake appears on the path. It does not help that we are wearing heavy boots, the fear is automatically there. We do not react in this way, when we bicycle in a street crowded with traffic and are exposed to real danger. The horror we feel when confronted with the snake, seems foolish and is out of proportion to the actual threat. Even a small, harmless spider in the window may cause the hair to stand on end due to panic.

Are there any explanations for these irrational responses? The possiblity that the fear, or phobia, has a genetic basis has been discussed. When the first human-like primates appeared in Africa about 4 million years ago, their number was small, whereas dangerous animals, like snakes, spiders, and scorpions, probably were numerous and far exceeded their number today. The human beings who most quickly learned to be afraid of the dangerous animals, also became the most skilfull in avoiding the danger and passed this quality on to coming generations. There are no reasons to believe that some of us are born with a fear of snakes. On the other hand, we may be born with a biological preparedness to more easily learn to be specifically afraid of a snake. It is known that if a parent is afraid of both city traffic and snakes, only the latter fear is easily tranferred to the child. The unreasonable fear of some venomous animals could therefore reflect something in our genetic makeup, which was of great significance for the survival of primitive man.

Bibliography

References of general and review character

Bakus, G.J. (1981). Chemical defense mechanisms on the Great Barrier Reef, Australia. Science, 211, 497–498.

Baslow, M.H. (1969). Marine Pharmacology. Williams and Wilkins Co., Baltimore.

Brower, L.P. (1969). Ecological chemistry. Scient. Amer., 220, (2), 22–29.

Bücherl, W., Buckley, E.E. and Deulofeu, V. (eds.). (1968–1971). Venomous Animals and their Venoms. Vols. I—III. Academic Press, Inc., New York.

Castle, N.A., Haylett, D.G. and Jenkinson, D.H. (1989). Toxins in the characterization of potassium channels. TINS, 12, (2), 59–65.

Cleland, J.B. and Southcott, R.V. (1965). Injuries to Man from Marine Invertebrates in the Australian Region. Commonwealth of Australia, Canberra, p. 195.

Eaker, D. and Wadström, T. (eds.). (1980). Natural toxins. Pergamon Press, Elmsford, N.Y.

Gans, C. (ed.). (1978). GIOLOGY OF THE REPTILIA. Vol. 8. Academic Press, London and New York.

Habermehl, G.G. (1981). Venomous Animals and Their Toxins. Springer-Verlag, Berlin.

Halstead, B.W. (1959). Dangerous marine animals. Cornell Maritime Press, Cambridge, Maryland.

Halstead, B.W. (1970). Venomous marine animals of the world. United States Government Printing Office, Washington, D.C.

Halstead, B.W. (1981). Current status of marine biotoxicology - an overview. Clin. Tox., 18, (1), 1–24.

Hashimoto, Y. (1979). Marine Toxins and Other Bioactive Metabolites. Japan Scientific Society, Tokyo.

Ivanov, Ch.P. and Ivanov, O.Ch. (1979). The evolution and ancestors of toxic proteins. Toxicon, 17, 205–220.

Ivanov, O.Ch. (1981). The evolutionary origin of toxic proteins. Toxicon 19, 171–178.

Kaul, P.N. and Daftari, P. (1986). Marine pharmacology: bioactive molecules from the sea. Ann. Rev. Pharmacol. Toxicol., 26, 117–142.

Kizer, K.W. (1984) Marine envenomations. J. Toxicol.-Clin. Toxicol., 21, (4 & 5), 527–555.

Longenecker, G.L. and Longenecker, H.E. (1984). Interactions of venoms and venom components with blood platelets. J. Toxicol.-Toxin Reviews, 3, (2 & 3), 223–251.

Meier, J. and Stocker, K. (1989). Review article. On the significance of animal experiments in toxinology. Toxicon, 27, (1), 91–104.

Parnas, I. and Russell, F.E. (1967). Effects of venoms on nerve, muscle and neuromuscular junction. In Russell F.E. and Saunders P.R. (eds.) Animal Toxins. Pergamon Press. Oxford, pp. 401–415.

Russell, F.E. (1963). Venomous animals and their toxins. London Times Sci, Rev., 49, 10.

Russell, F.E. (1967). Comparative pharmacology of some animal toxins. Fed. Proc., 26, 1206.

Russell, F.E. (1969). Posions and venoms. In Hoar, W.S. and Randall, D.J. (eds.) FISH PHYSIOLOGY. Vol. III. Academic Press, New York and London. pp. 401–449.

Russell, F.E. (1975). Poisonous and venomous marine animals and their toxins. Ann. N.Y. Acad. Sci., 245, 57–64.

Russell, F.E. (1986). Toxic effects of animals toxins. In Casarett and Doull's Toxicology. The basic science of poisons. (Klaassen C.D., Amdur, M.O. and Doull, J. eds.) Macmillan Publishing Company. New York, Toronto, London. pp. 706–755.

Russell, F.E. and Saunders, P.R. (eds.) (1967). Animal toxins. Pergamon Press, New York.

Schweitz, H. (1984). Lethal potency in mice of toxins from scorpion, Sea Anemone, snake and bee venoms following intraperitoneal and intracisternal injection. Toxicon, 22, (2), 308–311.

Shier, W.T. (1983). Toxins as research tools: potentials and pitfalls. J. Toxicol.-Toxin Reviews, 2, (1), 111–115.

Shier, W.T. (1984). Toxins as drugs: What does the future hold? J. Toxicol.-Roxin Reviews, 3, (2–3), 253–258.

Sutherland, S.K. (1981). Venomous creatures of Australia. A field guide with notes on first aid. Oxford University Press.

Tu, A.T. (ed.). (1984). HANDBOOK OF NATURAL TOXINS. Vol. 2. Insect Poisons, Allergens, and Other Invertebrate Venoms. Marcel Dekker, Inc., New-York-Basel.

Sponges, Corals and Sea anemones

Bernheimer, A.W., Avigad, L.S. and Kim, K.S. (1979). Comparison of metridiolysin from the sea anemone with thiol-activated cytolysins from bacteria. Toxicon, 17, 69–75.

Bernheimer, A.W., Avigad, L.S., Branch, G., Dowdle, E. and Lai, C.Y. (1984). Toxicon, 22, (2), 183–191.

Castle, N.A. and Strichartz, G.R. (1988). Palytoxin induces a relatively non-selective cation permeability in frog sciatic nerve which can be inhibited by cardiac glycosides. Toxicon, 26, (10), 941–951.

Culver, P. and Jacobs, R.S. (1981). Lophotoxin: a neuromuscular acting toxin from the sea whip (*Lophogorgia Rigida*). Toxicon, 19, (6), 825–830.

Doyle, J.W., Ken, W.R. and Vilallonga, F.A. (1989). Interfacial activity of an ion channel-generating protein cytolysin from the sea anemone *Stichodactyla Helianthus*. Toxicon, 27, (4), 465–471.

Habermann, E., Hudel, M. and Dauzenroth, M.-E. (1989). Palytoxin promotes potassium outflow from erythrocytes, HeLa and bovine adrenomedullary cells through its interaction with Na^+, K^+-ATPase. Toxicon, 27, (4), 419–430.

Habermann, E. (1989). Review article. Palytoxin acts through Na^+, K^+-ATPase. Toxicon, 27, (11), 1171–1187.

Lafranconi, W.M., Ferlan, I., Russell, F.E. and Huxtable, R.J. (1984). The action of equinatoxin, a peptide from the venom of the sea anemone, *Actinia Equina*, on the isolated lung. Toxicon, 22, (3), 347–352.

Mebs, D. and Gebauer, E. (1980). Isolation of proteinase inhibitory, toxic and hemolytic polypeptides from a sea anemone, *Stoichactis* SP. Toxicon, 18, 97–106.

Mebs, D., Liebrich, M., Reul, A. and Samejima, Y. (1983). Hemolysins and proteinase inhibitors from sea anemones of the Gulf of Aqaba. Toxicon, 21, (2), 257–264.

Moore, R.E. and Bartolini, G. (1981). Structure of parlytoxin. J. Am. Chem. Soc., 103, 2491–2494.

Norton, R.S., Bobek, G., Ivanov, J.O., Thomson, M., Fiala-Beer, E., Moritz, R.L. and Simpson, R.J. (1990). Purification and characterisation of proteins with cardiac stimulatory and haemolytic activity from the anemone *Actinia Tenebrosa*. Toxicon, 28, (1), 29–41.

Pichon, Y. (1982). Effects of palytoxin on sodium and potassium permeabilities in unmyelinated axons. Toxicon, 20, (1), 41–47.

Rack, M., Meves, H., Béress, L. and Grünhagen, H.H (1983). Preparation and properties of fluorescence labeled neuro- and cardiotoxin II from the sea anemone (*Anemonia Sulcata*). Toxicon, 21, (2), 231–237.

Ravens, U. and Schöllhorn, E. (1983). The effects of a toxin (ATX II) from the sea anemone *Anemonia Sulcata* on the electrical and mechanical activity of the denervated hemidiaphragm of the rat. Toxicon, 21, (1), 131–142.

Sevcik, C. and Barboza, C.A. (1983). The presynaptic effect of fractions isolated from the sponge *Tedania Ignis*. Toxicon, 21, (2), 191–200.

Hydroids, Jellyfishes and Polychaetes

Bon, C., Saliou, B., Thieffry, M. and Manaranche, R. (1985). Partial purification of a-glycerotoxin, a presynaptic neurotoxin from the venom glands of the polychaete annelid *Glycera Convoluta*. Neurochem. Int., 7, (1), 63–75.

Burnett, J.W. and Calton, G.J. (1977). Review article. The chemistry and toxicology of some venomous pelagic coelenterates. Toxicon, 15, 177–196.

Cobbs, C.S., Drzymala, R.E., Shamoo, A.E., Calton, G.J. and Burnett, J.W. (1983). Sea nettle (*Chrysaora Quinquecirrha*) lethal factor: effect on black lipid membranes. Toxicon, 21, (4), 558–561.

Kelman, S.N., Calton, G.J. and Burnett, J.W. (1984). Isolation and partial characterization of a lethal sea nettle (*Chrysaora Quinquecirrha*) mesenteric toxin. Toxicon, 22, (1), 139–144.

Klug, M., Weber, J. and Tardent, P. (1989). Hemolytic and toxic properties of *Hydra Attenuata* nematocysts. Toxicon, 27, (3), 325–339.

Lin, W.W., Lee, C.Y. and Burnett, J.W. (1988). Effect of sea nettle (*Chrysaora Quinquecirrha*) venom on isolated rat aorta. Toxicon, 26, (12), 1209–1212.

Neeman, I., Calton, G.J. and Burnett, J.W. (1980). Cytotoxicity and dermonecrosis of sea nettle (*Chrysaora Quinquecirrha*) venom. Toxicon, 18, 55–63.

Olson, C.E., Pockl, E.E., Calton, G.J. and Burnett, J.W. (1984). Immunochromatographic purification of a nematocyst toxin from the cnidarian *Chironex Fleckeri* (sea wasp). Toxicon, 22, (5), 733–742.

Shryock, J.C. and Bianchi, C.P. (1983). Sea nettle (*Chrysaora Quinquecirrha*) nematocyst venom: mechanism of action on muscle. Toxicon, 21, (1), 81–95.

Walker, M.J.A. (1977). The cardiac actions of a toxin-containing material from the jellyfish, *Cyanea Capillata*. Toxicon, 15, 15–27.

Gastropods and Octopods

Anthoni, U., Bohlin, L., Larsen, C., Nielsen, P., Nielsen, N.H. and Christophersen, C. (1989). The toxin tetramine from the "edible" whelk *Neptunea Antiqua*. Toxicon, 27, (7), 717–723.

Clark, C., Olivera, B.M. and Cruz, L.J. (1981). A toxin from the venom of the marine snail *Conus Geographus* which acts on the vertebrate central nervous system. Toxicon, 19, (5), 691–699.

Elliott, E.J. and Raftery, M.A. (1979). Venom of marine snail *Conus Californicus:* Biochemical studies of a cholinomimetic component. Toxicon, 17, 259–268.

Fänge, R. (1984). Venoms and venom glands of marine molluscs. In Bolis et al. (eds.). Toxins, drugs and pollutants in marine animals. Springer-Verlag, Berlin, Heidelberg, pp. 47–62.

Kobayashi, J., Ohizumi, Y., Nakamura, H. and Hirata Y. (1981). Pharmacological study on the venom of the marine snail *Conus Textile*. Toxicon, 19, (6), 757–762.

Noguchi, T., Maruyama, J., Narita, H. and Hashimoto, K. (1984). Occurrence of tetrodotoxin in the gastropod mollusk *Tutufa Lissostoma* (Frog shell). Toxicon, 22, (2), 219–226.

Olivera, B.M., Gray, W.R., Zeikus, R., McIntosh, J.M., Varga, J., de Santos, J.R.V. and Cruz, L.J. (1985). Peptide neurotoxins from fish-hunting cone snails. Science, 230, 1338–1343.

Olivera, B.M., Rivier, J., Clark, C., Ramilo, C.A., Corpuz, G.P., Abogadie, F.C., Mena, E.E., Woodward, S.R., Hillyard, D.R, and Cruz, L.J. (1990). Diversity of *Conus* neuropeptides. Science, 249, 257–263.

Sheumack, D.D., Howden, M.E.H., Spence, I. and Quinn, R.J. (1978). Maculotoxin: A neurotoxin from the venom glands of octopus *Hapalochlaena maculosa* identified as tetrodotoxin. Science, 199, 188–189.

Echinoderms

Alender, C.B. and Russell, F.E. (1966). Pharmacology, In Boolootin R.A. (ed.). Physiology of Echinodermata. Interscience, New York, p 529.

Anisimov, M.M., Shentsova, E.B., Shcheglov, V.V., Shumilov, Y.N., Rasskazov, V.A., Strigina, L.I., Chetyrina, N.S. and Elyakov, G.B. (1978). Mechanism of cytotoxic action of some triterpene glycosides. Toxicon, 16, 207–218.

Bakus, G.J. and Green, G. (1974). Toxicity in sponges and holothurians. Science, 185, 951–53.

Burnell, D.J. and ApSimon, J.W. (1983) Echinoderm saponins. In *Marine Natural Products,* ed. P.J. Scheuer, New York/London: Academic. 5, 365–69.

Gorshkov, B.A., Gorshkova, I.A., Stonik, V.A. and Elyakov, G.B. (1982). Short communications. Effect of marine glycosides on adenosinetriphosphatase activity. Toxicon, 20, (3), 655–658.

Kimura, A., Nakagawa, H., Hayashi, H. and Endo, K. (1984). Seasonal changes in contractile activity of a toxic substance from the pedicellaria of the sea urchin *Toxopneustes Pileolus.* Toxicon, 22, (3), 353–358.

Mackie, A.M., Singh, H.T. and Owen, J.M. (1977). Studies on the distribution, biosynthesis and function of steroidal saponins in echinoderms. Comp. Biochem. Physiol. B, 56, 9–14.

Parker, C.A. (1981). Poisonous qualities of the starfish. Zoologist, 5, 214–215.

Ruggieri, G.D. and Nigrelli, R.F. (1974). Physiologically active substances from echinoderms. In *Bioactive compounds from the Sea,* ed., Humm, H. and Lane, C. New York: Dekker. pp. 183–95.

Insects

Bettini, S. (ed.). (1978). Arthropod Venoms. Handbook of Experimental Pharmacoloby. Vol. 48. Springer-Verlag, Berlin, Heidelberg, New York.

Blum, M.S. (1981). Chemical defenses of arthropods. Adacemic Press.

Blum, M.S. (1984). Poisonous Ants and Their Venoms. In Tu A.T. (ed.) HANDBOOK OF NATURAL TOXINS. Vol. 2. Insect Poisons, Allergens and Other Invertebrate Venoms. Marcel Dekker, Inc. New York - Basel. pp. 225–242.

Cavey, D., Vincent, J.-P. and Lazdunski, M. (1979). A search for the apamin receptor in the central nervous system. Toxicon, 17, 176–179.

Cook, B.J. and Holman, G.M. (1985). Peptides and Kinins. In Kerkut G.A. and Gilbert L.I. (eds.). Comprehensive Insect Physiology Biochemistry And Pharmacoloby. Vol. 11. Pergamon Press, Oxford. pp. 531–593.

Einarsson, R. and Renck, B. (1984). Ion-exchange chromatographic characterization of stinging insect vespid venoms. Toxicon, 22 (1), 154–160

Jacobsen, T.F., Sand, O., Bjøro, T., Karlsen, H.E. and Iversen, J.-G. (1990). Effect of diamphidia toxin, a bushman arrow poison, on ionic permeability in nucleated cells. Toxicon, 28, (4), 435–444.

Kao, C.Y., Salwen, M.J., Hu, S.L., Pitter, H.M. and Woollard, J.M.R. (1989). Diamphidia toxin, the bushmen's arrow poison: possible mechanism of prey-killing. Toxicon, 27, (12), 1351–1366.

Morse, D.H. (1985). Milkweeds and their visitors. Scient. Amer. 253, (1), 90–96.

Ouyang, C., Lin, S.-C., and Teng, C.-M. (1979). Anticoagulant properties of *Apis Mellifera* (honey bee) venom. Toxicon, 17, 197–202.

Owen, M.D. and Bridges, A.R, (1974). The sting of the honey bee. Sci. Terrapin, 4, 9–12.

Piek, T. (1985). Insect Venoms and Toxins. In Kerkut G.A. and Gilbert L.I. (eds.) COMPREHENSIVE INSECT PHYSIOLOGY BIOCHEMISTRY AND PHARMACOLOGY, Vol. II, Pergamon Press, pp. 595–633.

Piek, T. (1984). Pharmacology of Hymenoptera Venoms. In Tu A.T. (ed.). HANDBOOK OF NATURAL TOXINS. Vol. 2. Insect Poisons, Allergens, and Other Invertebrate Venoms. Marcel Dekker, Inc. New York - Basel. pp. 135–185.

Prestwich, G.D. (1983). The chemical defenses of termites. Scient. Amer. 249, (2), 68–75.

Prince, R.C., Gunson, D.E. and Scarpa, A. (1985). Sting like a bee! The ionophoric properties of melittin. Trends in Biochemical Sciences, 10, (3), 99.

Rosenthal, G.A. (1986). The chemical defenses of higher plants. Scient. Amer., 254, (1), 76–81.

Schmidt, J.O., Blum, M.S. and Overal, W.L. (1980). Comparative lethality of venoms from stinging hymenoptera. Toxicon, 18, 469–474.

Woollard, J.M.R., Fuhrman, F.A. and Mosher, H.S. (1984). The bushman arrow toxin, diamphidia toxin: Isolation from pupae of *Diamphidia Nigro-Ornata.* Toxicon, 22, (6), 937–946.

Zlotkin, E. (1985). Toxins derived from arthropod venoms specifically affecting insects. In Kerkut G.A. and Gilbert L.I. (eds.). COMPREHENSIVE INSECT PHYSIOLOGY BIOCHEMISTRY AND PHARMACOLOGY. Vol. 10. Pergamon Press, Oxford. pp. 499–546.

Spiders and Scorpions

Babcock, J.L. Civello, D.J. and Geren, C.R. (1981). Purification and characterization of a toxin from brown recluse spider (*Loxosceles Reclusa*) venom bland extracts. Toxicon, 19, (5), 677–689.

Bablito, J., Jover, E. and Couraud, F. (1986). Activation of the voltage-sensitive sodium channel by a β-scorpion toxin in rat brain nerve-ending particles. J. Neurochem., 1763–1770.

Bachmann, M. (1982). Isolation and partial characterization of a toxin from the venom of the east african orthognath spider *Pterinochilus spec.* Toxicon, 20, (3), 547–552.

Brook, G., Cruz-Höfling, M.A., Duchen, L.W. and Love, S. (1984). Effects of the venom of the spider *Phoneutria nigriventer* on peripheral nerve and neuromuscular transmission in the mouse. J. Physiol., 360, 41P.

Cruz-Höfling, M.A., Love, S., Brook, G. and Duchen, L.W. (1985). Effects of *Phoneutria Nigriventer* spider venom on mouse peripheral nerve. Quarterly J. Exp. Physiol., 70, 623–640.

Dehesa-Dávila, M. (1989). Epidemiological characteristics of scorpion sting in León, Guanajuato, México. Toxicon, 27, (3), 281–286.

Foil, L.D., Frazier, J.L. and Norment, B.R. (1979). Partial characterization of lethal and neuroactive components of the brown recluse spider (*Loxosceles Reclusa*) venom. Toxicon, 17, 347–354.

Grishin, E.V., Volkova, T.M. and Arseniev, A.S. (1989). Isolation and structure analysis of components from venom of the spider *Argiope Lobata*. Toxicon, 27, (5), 541–549.

Kawai, N., Miwa, A. and Abe, T. (1983). Specific antagonism of the glutamate receptotr by an extract from the venom of the spider *Araneus Ventricosus*. Toxicon, 21, (3), 438–440.

Kharrat, R., Darbon, H., Granier, C. and Rochat, H. (1990). Structure-activity relationships of scorpion a-neurotoxins: contribution of arginine residues. Toxicon, 28, (5), 509–523.

Love, S., Cruz-Höfling, M.A. and Duchen, L.W. (1986). Morphological abnormalities in myelinated nerve fibres caused by *Leiurus*, *Centruroides* and *Phoneutria* venoms and their prevention by tetrodotoxin. Quarterly J. Exp. Physiol., 71, 115-122.

Lucas, S. (1988). Spiders in Brazil. Toxicon, 26, (9), 759-772.

Martin, M.F., Perez, L.G.G., El Ayeb, M., Kopeyan, C., Bechis, G., Jover, E. and Rochat, H. (1987). Purification and chemical and biological characterizations of seven toxins from the mexican scorpion, *Centruroides suffusus suffusus*. J. Biol. Chem., 262, (10), 4452–4459.

Meves, H., Simard, J.M. and Watt, D.D. (1984). Biochemical and electrophysiological characteristics of toxins isolated from the venom of the scorpion *Centruroides sculpturatus*. J. Physiol., Paris, 79, 185–191.

Mylecharane, E.J., Spence, I. and Gergson, R.P. (1984). In vivo actions of atraxin, a protein neurotoxin from the venom glands of the funnel-web spider (*Atrax Robustus*). Comp. Biochem. Physiol., 79C, (2), 395–399.

Mylecharane, E.J., Spence, I., Sheumack, D.D., Claassens, R. and Howden, M.E.H. (1989). Actions of robustoxin, a neurotoxic polypeptide from the venom of the male funnel-web spider (*Alrax Robustus*), in anaesthetized monkeys. Toxicon, 27, (4), 481–492.

Ribeiro, L.A., Jorge, M.T., Piesco, R.V. and Nishioka, S.A. (1990). Short communications. Wolf spider bites in Sào Paulo, Brazil: a clinical and epidemiological study of 515 cases. Toxicon, 28, (6), 715–717.

Romey, G., Abita, J.P., Chicheportiche, R., Rochat, H. and Lazdunski, M. (1976). Scorpion neurotoxin. Mode of action on neuromuscular junctions and synaptosomes. Biochim. Biophys. Acta, 448, 607–619.

Schenone, H. and Suarez, G. (1978). Venoms of Scytodidae, Genus Loxosceles. In Bettini S. (ed.). Arthropod Venoms. Springer-Verlag. New York, pp. 247–275.

Simard, J.M., Meves, H. and Watt, D.D. (1986). Effects of toxins VI and VII from the scorpion *Centruroides sculpturatus* on the Na currents of the frog node of Ranvier. Pflügers Arch., 406, 620–628.

Terakawa, S., Kimura, Y., Hsu, K. and Ji, Y.-H. (1989). Lack of effect of a neurotoxin from the scorpion *Buthus Martensi* Karsch on nerve fibers of this scorpion. Toxicon, 27, (5), 569–578.

Usmanov, P.B., Kalikulov, D., Shadyeva, N.G., Nenilin, A.B. and Tashmukhamedov, B.A. (1985). Postsynaptic blocking of glutamatergic and cholinergic synapses as a common property of araneidae spider venoms. Toxicon, 23, (3), 528–531.

Vijverberg, H.P.M. and Lazdunski, M. (1984). A new scorpion toxin with a very high affinity for sodium channels, an electrophysiological study. J. Physiol., Paris, 79, 275–279.

Watt, D.D, and Simard, J.M. (1984). Neurotoxic proteins in scorpion venom. J. Toxicol.-Toxin Reviews, 3, (2–3), 181–222.

Fishes

Arnold, S.H. and Brown, W.D. (1978). Histamine toxicity from fish products. Adv. Food Res., 24, 113.

Banner, A.H. (1976). Ciguatera: a disease from coral reef fish. In Jones O.E. and Endean R. (eds.). Biology and geology of coral reefs. Vol. III. Academic Press, Inc., New York, p. 177.

Benedek, C. and Rivier, L. (1989). Evidence for the presence of tetrodotoxin in a powder used in Haiti for zombification. Toxicon, 27, (4), 473–480.

Benoit, E., Legrand, A.M. and Dubois, J.M. (1986). Effects of ciguatoxin on current and voltage clamped frog myelinated nerve fibre. Toxicon, 24, (4), 357–364.

Cain, D. (1983). Weever fish sting: an unusual problem. British Med. J. 287, 406–407.

Cameron, A.M., Surridge, J., Stablum, W. and Lewis, R.J. (1981). A crinotoxin from the skin tubercle glands of a stonefish (*Synanceia Trachynis*). Toxicon, 19, 159–170.

Clark, E. (1974). The Red sea's sharkproof fish. National Geographic, 718–727.

Evans, M.H. (1967). Block of sensory nerve conduction in the cat by mussel poison and tetrodotoxin. In Russell F.E. and Saunders P.R. (eds.). Aninal Toxins. Pergamon Press, Oxford, p. 97.

Goodger, W.P. and Burns, T.A. (1980). The cardiotoxic effects of alligator gar (*Lepisosteus spatula*) roe on the isolated turtle heart. Toxicon, 18, 489–494.

Hori, K., Fusetani, N., Hashimoto, K., Aida, K. and Randall, J.E. (1979). Occurrence of a grammistin-like mucous toxin in the clingfish *Diademichthys Lineatus*. Toxicon, 17, 418–424.

Huang, J.M.C., Wu, C.H. and Baden, D.G. (1984). Depolarizing action of a red-tide dinoflagellate brevetoxin on axonal membranes. J. Pharmacol. Exp. Therapeut., 229, (2), 615–621.

Kodama, M., Ogata, T., Noguchi, T., Maruyama, J. and Hashimoto, K. (1983). Occurrence of saxitoxin and other toxins in the liver of the pufferfish *Takifugu Pardalis.* Toxicon, 21, (6), 897–900.

Kodama, A.M. and Hokama, Y. (1989). Variations in symptomatology of ciguatera poisoning. Toxicon, 27, (5), 593–595.

Legrand, A.M. and Bagnis, R. Short communications. Effects of ciguatoxin and maitotoxin on isolated rat atria and rabbit duodenum. Toxicon, 22, (3), 471–475.

Lewis, R.J. and Endean, R. Ciguatoxin from the flesh and viscera of the barracuda, *Sphyraena Jello.* Toxicon, 22, (5), 805–810.

Nakamura, M., Oshima, Y. and Yasumoto, T. (1984). Occurrence of saxitoxin in puffer fish. Toxicon, 22, (3), 381–385.

Nukina, M., Koyanagai, L.M. and Scheuer, P.J. (1984). Two interchangeable forms of ciguatoxin. Toxicon, 22, (2), 169–176.

Perriere, C., Goudey-Perriere, F. and Petek, F. (1988). Purification of a lethal fraction from the venom of the weever fish, *Trachinus Vipera* C.V. Toxicon, 26, (12), 1222–1227.

Primor, N. and Lazarovici, P. (1981). *Pardachirus Marmoratus* (Red Sea flatfish) secretion and its isolated toxic fraction pardaxin: the relationship between hemolysis and ATPase inhibition. Toxicon, 19, (4), 573–578.

Russell, F.E. and Emery, J.A. (1960). Venom of the weevers Trachinus draco and Trachinus vipera. Ann. N.Y. Acad. Sci., 90, 805.

Thompson, S.A., Tachibana, K., Nakanishi, K. and Kubota, I. (1986). Melittin-like peptides from the shark-repelling defense secretion of the sole *Pardachirus pavoninus.* Science, 233, 341–343.

Vietmeyer, N.D. (1984). The Preposterous Puffer. National Geographic, 260–270.

Wu, C.H., Huang, J.M.C., Vogel, S.M., Luke, V.S., Atchison, W.D. and Narahashi, T. (1985). Actions of *Ptychodiscus Brevis* toxins on nerve and muscle membranes. Toxicon, 23, (3), 481–487.

Amphibians

Bevins, C.L. and Zasloff, M. (1990). Peptides from frog skin. Ann. Rev. Biochem., 59, 395–414.

Brandon, R.A. and Huheey, J.E. (1981). Toxicity in the plethodontid salamanders *Pseudotriton Ruber* and *Pseudotriton Montanus* (Amphibia, Caudata). Toxicon, 19, 25–31.

Daly, J.W. (1982). Biologically active alkaloids from poison frogs (Dendrobatidae). Toxin Rev., 1, 33.

Daly, J.W., Highet, R.J. and Myers, C.W. (1984). Occurrence of skin alkaloids in non-dendrobatid frogs from Brazil (bufonidae), Australia (myobatrachidae) and Madagascar (mantellinae). Toxicon, 22, (6), 905–919.

Flier, J., Edwards, M.W., Daly, J.W. and Myers, C.W. (1980). Widespread occurrence in frogs and toads of skin compounds interacting with the ouabain site of Na^+, K^+-ATPase. Science, 208, 503–505.

Huheey, J.E. and Brandon, R.A. (1977). Novel toxins and the question of warning coloration and mimicry in salamanders. Herpetol. Rev., 8, Suppl. 10.

Jaeger, R.G. (1971). Toxic reactions to skin secretions of the frog. *Phryonomerus bifasciatus.* Copeia, 1971, (1), 160–161.

Myers, C.W. and Daly, J.W. (1983). Dart-poison frogs. Scientific American, 248, (2), 97–105.

Nakagima, T. (1981). Active peptides in amphibian skin. Trends Pharmac. Sci. 2, 202–205.

Otani, A., Palumbo, N. and Read, G. (1969). Pharmacodynamics and treatment of mammals poisoned by *Bufo marinus* toxin. Am. J. Vet. Res., 30, 1865–72.

Tokuyama, T., Daly, J.W. and Highet, R.J. (1984). Pumiliotoxins: magnetic resonance spectral assignments and structural definition of pumiliotoxins A and B and related allopumiliotoxins. Tetrahedron, 40, 1183–90.

Reptiles

Allen, M. and Tu, A.T. (1985). The effect of tryptophan modification on the structure and function of a sea snake neurotoxin. Mol. Pharmacol., 27, 79–85.

Benoit, E. and Dubois, J.-M. (1986). Toxin I from the snake *Dendroaspis polylepis polylepis:* a highly specific blocker of one type of potassium channel in myelinated nerve fiber. Brain Res., 377, 374–377.

DaSilva, N.J., Aird, S.D., Seebart, C. and Kaiser, I.I. (1989). A gyroxin analog from the venom of the bushmaster (*Lachesis muta muta*). Toxicon, 27, (7), 763–771.

de Wit, C.A. (1987). The hedgehog's (Erinaceus europaeus) natural resistance to european viper (Vipera berus) venom: The role of β-macroglobulin inhibitor as antihemorrhagic factor. Doctoral dissertation, Lund University, Dept of Zoophysiology.

Elliott, W.B. (1978). Chemistry and immunology of reptilian venoms. In Gans C. (ed.). Biology of the Reptilia. Vol. 8. Academic Press, London. pp. 163–436.

Fraenkel-Conrat, H. (1982–83). Snake venom neurotoxins related to phospholipase A2. J. Toxicol.-Toxin Reviews, 1, (2), 205–221.

Garcia, V.E. and Perez, J.C. (1984). The purification and characterization of an antihemorrhagic factor in woodrat (*Neotoma Micropus*) serum. Toxicon, 22, (1), 129–138.

Glenn, J.L., Straight, R.C., Wolfe, M.C. and Hardy, D.L. (1983). Geographical variation in *Crotalus scutulatus Scutulatus* (Mojave rattlesnake) venom properties. Toxicon, 21, (1), 119–130.

Harvey, A.L., Anderson, A.J., Mbugua, P.M. and Karlsson, E. (1984). Toxins from mamba venoms that facilitate neuromuscular transmission. J. Toxicol.-Toxin Reviews, 3, (2 & 3), 91–137.

Harvey, A.L. (1985). Carciotoxins from cobra venoms: possible mechanisms of action. J. Toxicol.-Toxin reviews, 4, (1), 41–69.

Hayashi, K., Endo, T., Nakanishi, M., Furukawa, S., Jorbert, F.J, Nagaki, Y., Nomoto, H. and Tamiya, N. (1986). On the mode of action of snake postsynaptic neurotoxins. J. Toxicol.-Toxin Reviews, 5, 92), 95–104.

Ho, C.L., Tsai, I.H. and Lee, C.Y. (1986). The role of enzyme activity and charge properties on the presynaptic neurotoxicity and the contracture-inducing activity of snake venom phospholipases A_2. Toxicon, 24, (4), 337–345.

Huang, S.-Y. and Perez, J.C. (1980). Comparative study on hemorrhagic and proteolytic activities of snake venoms. Toxicon, 18, 421–426.

Huang, S.Y. and Perez, J.C. (1980). Comparative study on hemorrhagic and proteolytic activities of snake venoms. Toxicon, 18, 421–426.

Ishizaki, H., Allen, M. and Tu, A.T. (1983). Effect of sulfhydryl group modification on the neurotoxic action of a sea snake toxin. J. Pharm. Pharmacol. 36, 36–41.

Kasckow, J.W., Abood, L.G., Hoss, W. and Herndon, R.M. (1986). Mechanism of phospholipase A_2-induced conduction block in bullfrog sciatic nerve. I. Electrophysiology and morphology. Brain Res., 373, 384–391.

Kornalik, F. (1985). The influence of snake venom enzymes on blood coagulation. Pharmac. Ther., 29, 353–405.

Lee, C.-Y (ed.) (1979). Snake Venoms. Handbook of Experimental Pharmacology. Vol. 52. Springer-Verlag. Berlin, Heidelberg, New York.

Markland, F.S. (1983). Rattlesnake venom enzymes that interact with components of the hemostatic system. J. Toxicol.-Toxin Reviews, 2, (2), 119–160.

Mebs, D. (1978). Pharmacology of Reptilian Venoms. In Gans C. (ed.) BIOLOGY OF THE REPTILIA. Vol. 8. Academic Press. London. pp. 437–560.

Ménez, A., Boulain, J.-C., Bouet, F., Couderc, J., Faure, G., Rousselet, A., Trémeau, O., Gatineau, É. and Fromageot, P.

(1984). On the molecular mechanisms of neutralization of a cobra neurotoxin by specific antibodies. J. Physiol., Paris, 79, 196–206.

Minton, S.A. Jr. and Minton, M.G. (1969). Venomous Reptilies. Charles Scribner's Sons. New York.

Mochca-Morales, J., Martin, B.M. and Possani, L.D. (1990). Isolation and characterization of helothermine, a novel toxin from *Heloderma Horridum Horridum* (Mexican beaded lizard) venom. Toxicon, 28, (3), 299–309.

Mors, W.B. DoNascimento, M.C., Parente, J.P., DaSilva, M.H., Melo, P.A. and Suarez-Kurtz, G. (1989). Neutralization of lethal and myotoxic activities of South American rattlesnake venom by extracts and constituents of the plant *Eclipta Prostrata* (asteraceae). Toxicon, 27, (9), 1003–1009.

Radvanyi, F. and Bon, C. (1984). Investigations on the mechanism of action of crotoxin. J. Physiol., Paris, 79, 327–333.

Rapuano, B.E., Yang, C.-C and Rosenberg, P. (1986). The relationship between high-affinity noncatalytic binding of snake venom phospholipases A_2 to brain synaptic plasma membranes and their central lethal potencies. Biochim. Biophys. Acta, 856, 457–470.

Rugolo, M., Dolly, J.O. and Nicholls, D.G. (1986). The mechanism of action of β-bungarotoxin at the presynaptic plasma membrane. Biochem. J., 233, 519–523.

Russell, F.E. and Bogert, C.M. (1981). Gila monster: its biology, venom and bite - a review. Toxicon, 19, 341.

Selistre, H.S., Giglio, J.R. (1987). Isolation and characterization of a thrombin-like enzyme from the venom of the snake Bothrops insularis (Jararaca ilhoa). Toxicon, 25, 1135–1144.

Thomas, R.G. and Pough, F.H. (1979). The effect of rattlesnake venom on digestion of prey. Toxicon, 17, 221–228.

Tomihara, Y., Kawamura, Y., Yonaha, Nozaki, M., Yamakawa, M. and Yoshida, C. (1990). Neutralization of hemorrhagic snake venoms by sera of *Trimeresurus Flavoviridis* (Habu), *Her-*

pestes Edwardsii (Mongoose) and *Dinodon Senicarinatus* (Akamata). Toxicon, 28, (8), 989–991.

Tu, A.T. (1982). Rattlesnake venoms. Marcel Dekker, Inc., New York - Basel.

Tu, A.T. (1983). Local tissue damaging (hemorrhage and myonecrosis) toxins from rattlesnake and other pit viper venoms. J. Toxicol.-Toxin Reviews, 2, (2), 205–234.

Tu, A.T. and Hendon, R.R. (1983). Characterization of lizard venom hyaluronidase and evidence for its action as a spreading factor. Comp. Biochem. Physiol., 76B, (2), 377–383.

Wetzel, W.W. and Christy, N.P. (1989). A king cobra bite in New York city. Toxicon, 27, (3), 393–395.

Mammals

Calaby, J.H. (1968). The platypus (*Ornithorhynchus anatinus*) and its venomous characteristics. Vol. I, pp. 15–29. In *Venomous Animals and their Venoms.* (Bücherl, W., Buckley, E.E. and Deulofeu, V. eds.). Academic Press, Inc., New York.

Carson, K.A. and Rose, R.K. (1985). Ultrastructure of the submandibular gland of the venomous short-tailed shrew *Blarina - Carolinensis.* Anat. Rec., 211, (3), 35 A-36A.

Martin, I. G. (1981). Venom of the short-tailed shrew (*Blarina Brevicauda*) as an insect immobilizing agent. J. Mamm., 62, (1), 189–192.

Pournelle, G.H. (1986). Classification, biology and description of the venom apparatus of insectivores of the genera *Solenodon, Neomys,* and *Blarina.* Vol. I, pp. 31–42. In *Venomous Animals and their Venoms.* (Bücherl, W., E.E. and Deulofeu, V. eds.). Academic Press, Inc., New York.

Pucek, M. (1986). Chemistry and Pharmacology of insectivore venoms. Vol. I, pp. 43–50. In *Venomous Animals and their*

Venoms. (Bücherl, W., Buckley, E.E. and Deulofeu, V. eds.). Academic Press, Inc., New York.

Tomasi, T.E. (1978). Function of venom in the short-tailed shrew, *Blarina Brevicauda*. J. Mamm., 59, (4), 852–854.

Glossary

Acetylcholine, ACh, an acetic acid ester of choline, $CH_3COOCH_2CH_2N(CH_3)_3OH$, important neurotransmitter.

Acetylcholinesterase, enzyme that hydrolyzes acetylcholine.

ACh, see acetylcholine.

Action potential, electric signal propagated over long distances by nerve and muscle cells; characterized by transient all-or-none reversal of a membrane potential in which the inside of the cell temporarily becomes positive relative to the outside.

Activation of sodium channels, an increased conductance of excitable membranes to sodium ions in response to membrane depolarization. Depolarization opens sodium channels.

Adrenaline, also called epinephrine; hormone secreted by the adrenal medulla.

Affinity, the strength with which a ligand binds to its binding site.

Alcaloids, a group of naturally occurring organic nitrogenous bases; often pharmaceutically active.

Allergen, substance causing immune reactivity.

Alveoli, small cavities, the functional units of the lung.

Amino acid sequence, mutual order of amino acids in a protein.

Amphiphilic substance, molecule bearing both hydrophilic and hydrophobic groups.

Anaphylaxis, severe allergic reaction.

Antibody, an immunoglobulin protein that interacts only with the antigen that brought about its production.

Anticoagulant, prevents blood coagulation.

Antigen, a substance capable of bringing about the production of antibodies and then to react with them specifically.

Antivenin, antibody preparation to abolish toxic action of venom, poison or toxin antigens.

Arrhythmia, any variation from the normal heartbeat rhythm.

Autoimmune disease, a disease caused by antibodies against its own cells or proteins.

Axon, elongated cylindrical process of a nerve cell along which action potentials are conducted.

Bioamines, low molecular compounds containing basic (NH_2-) groups. Often derived from amino acids and showing biological activity.

Blood-brain barrier, anatomical barriers that control the kinds of substances which enter the extracellular space of the brain.

Blood platelets, a cell fragment present in the blood; plays several roles in blood clotting.

Botulinum toxin, a toxin from Bacillus botulinus; causes muscular fatigue and paralysis due to abolition of acetylcholine release.

Bradycardia, a reduction in heart rate from the normal level.

Bradykinin, a hormone formed from a precursor normally circulating in the blood; a potent vasodilator.

Calcium channel, a membrane channel which opens for calcium when activated.

Cardiotoxic substance, toxic for the heart.

Cardiovascular system, the heart and blood vessels.

Carotene, vitamin A, found in visual pigments and plants (e.g., carrots).

Catalase, enzyme which splits hydrogen peroxide (H_2O_2) to H_2O and O_2.

Cation, a positively charged ion; attracted to a negatively charged electrode.

Cholinergic, pertaining to acetylcholine; a compound which acts like acetylcholine.

Coevolution, some species are during evolution intimately adapted to each other and can not exist isolated from each other.

Collagenase, an enzyme causing breakdown of collagen, which is a strong fibrous protein serving as a structural element.

Curare, South American arrow poison; blocks synaptic transmission at the motor endplate by competetive inhibition of acetylcholine receptors.

Cytolytic substance, causes dissolution of cells.

Cytolysis, dissolution of cells.

Cytoskeleton, filaments of various kinds in the cytoplasm of cells; associated with cell shape and movements.

Depolarization, change of membrane potential toward zero, so that the inside of the cell becomes less negative.

Deoxyribonuclease, enzyme causing breakdown of DNA.

Dinoflagellates, an order among protozoans; unicellular organisms.

Dopamine, a catecholamine neurotransmitter.

Endemic, confined to a given region, e.g., an island or country.

Enzyme, a protein with catalytic properties.

Excitable cell, ability to produce action potentials.

Evolution, cumulative change in the characteristics of populations or organisms, occurring in the course of successive generations related by descent.

Fibrin, a protein fragment resulting from the enzymatic cleavage of fibrinogen and having the ability to polymerize and turn blood into a solid gel (clot).

Fibrinogen, a plasma protein which is the precursor of fibrin.

Globular protein, protein of spherical shape.

Glycoprotein, protein linked to a carbohydrate.

Glycoside, compound consisting of a sugar (most frequently glucose) and one or more other substances. Widely distributed in plants.

Hematotoxic compound, causing damage of blood.

Hemolymph, the blood of invertebrates with open circulatory systems.

Hemolysis, disruption of blood corpuscles, erythrocytes.

Hemorrhage, bleeding.

Herbivorous animal, a plant eater.

Histamine, the base formed from histidine by decarboxylation; responsible for dilation of blood vessels, secreted mainly by mast cells.

Histidine, an essential basic amino acid.

Homeothermic animal, an animal that regulates its own internal temperature within a narrow range, regardless of the ambient temperature.

Homologous substance, of similar chemical structure.

Hyaluronidase, enzyme causing breakdown of hyaluronic acid, a constituent of the intercellular matrix.

Hydrogen peroxide, H_2O_2; a strong oxidant, highly destructive to various molecules.

Hydrophilic, having an affinity for water.

Hydrophobic, lacking an affinity for water.

5-hydroxytryptamine, see serotonin.

Hypotensive substance, decreases arterial blood pressure.

Inactivation of sodium channel, the closing of sodium channels at the peak of the action potential.

Insecticide, insect poison.

Intercellular matrix, structures between cells.

Intracisternal, within the space (subarachnoid space) between the arachnoid membrane and the inner covering (pia mater) of the brain.

Intraperitoneal, within the membrane that lines the abdominal and pelvic cavities.

Intravenous, within the vein.

In vitro, in an artificial solution outside the body.

In vivo, within the living organism or tissue.

Ionophore, carrier molecule that promotes the permeation of ions across a membrane.

Kallikrein, enzyme catalyzing the release of kinin from kininogen.

K^+-channel, a membrane channel which opens for potassium in response to depolarization.

Kinins, peptides split from kininogen in inflamed areas; facilitate the vascular changes associated with inflammation; may also stimulate pain receptors.

Lactate dehydrogenase, enzyme catalyzing oxidation of lactate to pyruvic acid

Lipase, enzyme catalyzing the breakdown of lipids to free fatty acids and monoglycerides.

Mandible, one of the pair of mouth-parts of insects, which usually does most of the work of biting and crushing food.

Mast cell, a connective-tissue cell similar to a basophil except that it does not circulate in the blood; local injury releases histamine from mast cells.

Medusa, free-swimming form of coelenterate, e.g., jellyfish.

Membrane potential, the voltage difference betwen the inside and outside of the cell due to separation of charge across the membrane.

Metabolites, constituents of chemical reactions that occur within a living organism.

Metalloproteinase, proteinase containing a metal ion.

Mimicry, protective similarity in appearance of one species of animal to another.

Mitochondria, rod-shaped intracellular organelle, which produces the energy rich molecule, ATP; site of Krebs cycle and oxidative-phosphorylation enzymes.

Molarity, number of moles of solutes in a liter of solution.

Monovalent, having a valence of one.

Monomer, a compound capable of combining in repeating units to form a dimer, trimer, or polymer.

Motor nerve, transmits signals from the central nervous system to peripheral targets, for instance skeletal muscle fibers.

Mucopolysaccharides, a group of acidic polysaccharides present in the extracellular matrix of connective tissues.

Myonecrotic, structural damage of myofibrils (muscle fibers).

Na^+-K^+-ATPase, a membrane enzyme associated with extrusion of intracellular sodium and uptake of extracellular potassium. Responsible for maintenance of membrane potential.

Na^+-channels, see sodium channel.

Necrotic, structural damage of tissue.

Nerve growth factor (NGF), a peptide which stimulates the growth and differentiation of neurons of the sympathetic nervous system.

Neuralgia, pain localized to certain distinct sensory nerves.

Neuromuscular transmission, transfer of signal from nerve to muscle.

Neurotoxic substance, exerts toxic effects on the nervous system.

Neurotransmitter, molecule released by a presynaptic nerve ending, that interacts with receptor molecules in the postsynaptic membrane and influences the electrical activity of the postsynaptic cell.

Nicotinic receptor, a class of acetylcholine receptors that are sensitive to nicotine.

nm, 10^{-9} meter.

Noradrenaline, norepinephrine, a catecholamine neurotransmitter released at most sympathetic postganglionic nerve endings, from the adrenal medulla, and in many regions of the central nervous system.

Octopamine, a molecule resembling noradrenaline; a putative neurotransmitter in certain lower animals.

Oxidative metabolism, the process by which energy derived from the reaction between hydrogen and oxygen is transferred to ATP during its formation from ADP and inorganic phosphate; occurs in mitochondria.

Oxidize, removal of electrons.

Parasympathetic nervous system, craniosacral part of autonomic nervous system. Most of its nerve cells release the neurotransmitter acetylcholine.

Pedipalp, second head appendage of arachnids; may be locomotor, used for seizing prey, sensory or modified in the male for fertilization.

Penicillinase, enzyme which catalyzes the breakdown of penicillin.

Peptide, a short polypeptide chain having less than 50 amino acids.

Polypeptide, polymer consisting of amino acid subunits joined in sequence by peptide bonds; peptides and proteins.

pH, the negative logarithm to the base 10 of the hydrogen-ion concentration of a solution. The pH decreases as the acidity increases.

Phospholipase, enzyme which catalyzes the breakdown of phospholipids.

Phospholipids, lipids containing phosphorous, fatty acids, glycerol, and a nitrogenous base. A major component of cell membranes.

Phylum, one of the major kind of groups used in classifying animals, e.g., phylum Chordata. Consists of one class or a number of similar classes.

Poikilothermic animal, an animal whose body temperature more or less follows the ambient temperature.

Polar substance, substance containing a number of polar bonds; the region of the molecule to which the electrons are drawn carries a negative charge and the region from which they are drawn carries a positive charge; soluble in water.

Polyamines, consist of putrescine, spermidine, and spermine. Exist in all cells and play a role in growth and proliferation of cells.

Polyp, sedentary form of coelenterate, e.g., hydra, sea anemone.

Postsynaptic, located distal to the synaptic cleft.

Presynaptic, located proximal to the synaptic cleft.

Propagation, conduction of action potentials.

Prostaglandins, natural fatty acids which are able to induce contraction in uterine and other smooth muscles, lower blood pressure, and modify the actions of some hormones.

Proteolytic enzyme, same as protease.

Protease, enzyme which splits proteins by hydrolysis of peptide bonds.

Quantal release, the concept that neurotransmitter is released in multiples of discrete packets, which represent individual or groups of transmitter containing vesicles.

Receptor, molecules situated on the outer surface of the cell membrane and that interact specifically with messenger molecules, such as transmitters and hormones.

Reduce, the addition of electrons to a substance.

Ribonuclease, enzyme which breaks the bonds between the building blocks of RNA.

Sarcoplasmic reticulum, a smooth, membrane-limited network surrounding each myofibril; site of storage and release of calcium ions.

Schwann cell, a nonneural cell which surrounds all peripheral nerves; forms the myelin sheath arround a myelinated axon in the peripheral nervous system.

Sensory nerves, transmit information, which has a conscious correlate, to the central nervous system.

Serotonin, 5-hydroxytryptamine, 5-HT; a neurotransmitter, $C_{10}H_{12}N_2O$.

Sfingomyelin, a lipid molecule specific for the nervous system.

Sodium channel, a membrane channel which opens for sodium when activated by depolarization.

Sodium-potassium pump, active extrusion of sodium from the cell and uptake of potassium into the cell at the expense of metabolic energy.

Subcutaneous, under the skin.

Steroid, a subclass of lipid molecules; the molecule consists of a skeleton of four interconnected carbon rings to which a few polar groups may be attached.

Synapse, a specialized junction between two nerve cells, or between a nerve cell and a muscle cell, where impulses in the presynaptic cell influence the activity of the postsynaptic cell.

Synaptic cleft, the space at a synapse which separates the nerve cells, or a nerve cell and a muscle cell.

Synergistic effect, aids and enhances the intended response.

Tachycardia, an increase in heart rate above normal level.

Taxonomy, study of the classification of organisms according to their resemblances and differences.

Taurine, a putative transmitter, $H_2NCH_2CH_2SO_2OH$.

Thrombin, an enzyme that catalyzes the conversion of fibrinogen to fibrin.

Transmitter, a chemical mediator released from a presynaptic ending, that produces a conductance change or other response in the postsynaptic cell.

Tyramine, a putative transmitter in lower animals. Resembles dopamine.

μg, 10^{-6} gram.

μl, 10^{-6} liters.

Vasodilation, an increase in the diameter of blood vessels.

Vasoconstriction, a decrease in the diameter of blood vessels.

Vasopressin, a peptide hormone, increases the blood pressure.

Vesicle, a small, membrane-bound intracellular organelle.

Index